MATERIALS SCIENCE AND TECHNOLOGIES

Additional books in this series can be found on Nova's website under the Series tab.

Additional E-books in this series can be found on Nova's website under the E-books tab.

MATERIALS SCIENCE AND TECHNOLOGIES

POLYSTYRENE: PROPERTIES, PERFORMANCE AND APPLICATIONS

JAMES E. GRAY
EDITOR

Nova Science Publishers, Inc.
New York

For permission to use material from this book please contact us:
Telephone 631-231-7269; Fax 631-231-8175
Web Site: http://www.novapublishers.com

LIBRARY OF CONGRESS CATALOGING-IN-PUBLICATION DATA

Polystyrene : properties, performance, and applications / editors, James E. Gray.
p. cm.
Includes index.
ISBN 978-1-61209-117-4 (hardcover)
1. Polystyrene. I. Gray, James E. (James Ehren), 1960-
TP1180.S7P75 2010
668.4'233--dc22
2010047082

Published by Nova Science Publishers, Inc. ✢ New York

CONTENTS

PREFACE

Polystyrene is a common thermoplastic polymer made from the aromatic monomer styrene with good formability. It is widely used in automotive, electrical and electronic connector systems. This book presents current research in the study of polystyrene, including polystyrene attrition during chemical engineering processes; the biodegradability of polystyrene; the role of adhesion mechanisms of particles on modified polypropylene; direct fluorination of polystyrene; the applications of polystyrene into various sensing platforms; electrospun polystyrene fibers and superhydrophobic surfaces and the synthesis of end-functionalized syndiotactic polystyrene.

Chapter 1 - The issue of polymers films and particles friction, damage and attrition attracts both scientific and technological interest. In fact, polymers friction and attrition are of pivotal importance in many technological domains and applications such as powder technology, processes in fluidized beds, materials handling and transport processing (e.g., screw, pneumatic and hydraulic convey) and biopolymers and polymers processing. Moreover, polymer damage (e.g., attrition) has harmful effects on product quality, on environmental and health safety (*source of toxic dust*) and on the process reliability because of changes in particle properties such as particle shape and size distribution. The present contribution will focus on the attrition analysis induced by the rub of hemispherical polystyrene particles against smooth and functionalized silicon walls. Several experimental and structural factors affecting attrition of polymeric particles such as the velocity, the applied normal force, polymer molecular weights, and the wall surface energy are discussed. A homemade device was used to carry out a single slide of the polystyrene hemispherical (*4 nm radius*) particle onto different silicon walls. Atomic force microscopy (AFM) was used so as to analyze the attrition appearance induced by the polymer damage. Nano-friction experiments were also achieved by AFM in lateral mode in order to distinguish the effect of interfacial interactions at the nanometric scale.

Chapter 2 - In protecting the environment from pollution by waste polymers, it is indispensable to make synthetic polymers biodegradable by modification of the chemical structure without severe damage of utility. Biodegradation of polystyrene is extremely difficult, but copolymer of styrene with *N*-benzyl-4-vinylpyridinium chloride, PSt-*co*-BVP(Cl), is degradable when treated with activated sludge in soil. Biodegradation of the copolymer that contained 49.1-mol % BVP(Cl) followed first order kinetics and half-life was 5.6 days when 1.0 or 0.5 g/kg of polymer sample was used. Biodegradation of the copolymer with reduced BVP(Cl) content did not follow first order kinetics under the conditions, but appeared to follow the kinetics when a sufficiently small amount of polymer sample was used. Under ultimate conditions, half-life of the copolymer that contained 10.6-mol % BVP(Cl) was estimated 12.5 days. Half-life of the copolymer that contained 5-mol % BVP(Cl) was assumed 30-40 days. Insertion of 5-mol % BVP(Cl) appeared sufficient for making the copolymer substantially biodegradable if the author do not expect exceptionally rapid degradation. Random scission of the main chain predominated over uniform scission from the end of polymer chain in the biodegradation of PSt-*co*-BVP(Cl). Cleavage of the main chain at BVP(Cl) appeared predominant over that at oligo-styrene portion. There are a variety of noticeable utilities of the modified polystyrene due to the strong affinity of BVP(Cl) with microbial cells and viruses.

Chapter 3 - The present work should be considered in the context of the health safety and hygiene of materials. In several areas of application of plastic materials (agro-food industry, medical devices or even pharmaceuticals) it is necessary to understand how a surface reacts in contact with the surrounding environment (food, living beings, etc.) so that surfaces can be developed that meet their requirements. The aim of this work was therefore to understand the role of surface parameters implicated in adhesion phenomena. The use of spherical particles (i.e. polystyrene or silica beads) as a model was designed to enable adhesion tests during which their surface properties would not vary.

The first challenge of this work was to modify the surface of polypropylene (PP) by radio-grafting with hydrophilic functional molecules such as N, N-dimethylacrylamide (DMA) and [2-methacryloyloxy)ethyl] trimethylammonium chloride, which is a quaternary ammonium salt (QAS). These monomers were used alone or in well defined mixtures. The aim of grafting was to modulate the surface parameters of PP such as wettability, roughness and surface charge.

Determination of the degree of grafting by weighing combined with infrared (FTIR-ATR) and X-Ray photoelectron spectroscopy (XPS) analysis of the surface confirmed the grafting of the molecules. Analysis of the physicochemical characteristics of the modified surfaces showed that grafting had changed these surface parameters as a function of the grafting conditions (concentration of grafting solution, type of molecule).

The second part of this work was designed to understand the mechanisms involved in adhesion phenomena. Particles adhering to the modified samples were studied according to the surface parameters. Adhesion tests confirmed the strong influence of substrate type (mainly hydrophilicity and roughness) and to a lesser extent underlined the role of electrostatic interactions.

Chapter 4 - In this paper, fundamental features of the direct fluorination of polystyrene is reviewed. Direct fluorination of polymers (i.e. treatment of a polymer surface with gaseous fluorine and its mixtures) proceeds at room temperature spontaneously and can be considered as a surface modification process. A large variety of experimental methods, such as FTIR spectroscopy, visible and near UV spectroscopy, Electron Spin Resonance spectroscopy, laser interference spectroscopy, refractometry, etc. was applied. Fundamental features of the direct fluorination, such as influence of treatment conditions (composition of the fluorinating mixture, fluorine partial pressure, fluorination duration) on the rate of formation, chemical composition, density, refraction index of the fluorinated layer, formation of radicals during fluorination and their termination were studied. On the base of obtained experimental data a simple theoretical model of the direct fluorination of polymers was developed.

Chapter 5 - Polymeric materials have gained a wide theoretical interest and practical application in sensor technology. Polymers offer a lot of advantages for sensor technologies: they are relatively low cost materials, their fabrication techniques are quite simple (there is no need for special clean-room and/or high temperature processes), they can be deposited on various types of substrates and the wide choice of their molecular structure and the possibility to build in side-chains, charged or neutral particles, and even grains of specific behavior into the bulk material or on its surface region, enables films to be produced with various physical and chemical properties. These properties can be adapted for sensing behavior.

Out of the various polymers currently available, polystyrene, because of its preferable physical, optical and electrical properties, has generated a considerable amount of interest in the sensor community. Polystyrene has found applications more specifically in the biological, chemical and optical sensor areas. The scope of this review chapter is to demonstrate the applications of polystyrene into various sensing platforms. The various physical and chemical forms in which polystyrene is used are elaborated. The various detection methods (electrical, optical, mass based) are also elaborated. Finally, the present status and perspectives as well as the advantages of the specific polystyrene based sensors are summarized.

Chapter 6 - Polystyrene (PS), a common thermoplastic polymer made from the aromatic monomer styrene with good formability, is widely used in automotive, electrical and electronic connector systems. During the past decade, electrospinning as a versatile technique for continuous fabrication of microscale- or nanoscale-diameter fibers has attracted tremendous interests for the fabrication of ultrathin PS fibers. In this chapter, PS fibers with various morphologies including uniform fibers and beaded fibers were prepared via regulating the solution properties. Furthermore, PS fibers with the combination of micro- and nanostructured surface roughness inherently based on the inspiration of self-cleaning silver ragwort leaves were achieved directly using mixed solvents. The pure PS fibrous mats showed the superhydrophobicity with a water contact angle (WCA) of 154° and a water contact angle hysteresis of 6°. However, the practical applications for these superhydrophobic mats were significantly limited due to their inadequacy of mechanical integrity. To enhance the mechanical properties of PS mats, polyamide 6 and polyacrylonitrile fibers were blended into the PS mats via multi-syringe electrospinning technique, respectively. At the WCA of 150°, the PS fibrous mats blended with added-fibers showed a 3 times increase in the tensile strength compared with the pure PS fibrous mats.

Chapter 7 - The use of syndiotactic polystyrene for practical applications can be hampered by its brittleness, lack of adhesion and processability at high temperatures. These properties can be improved by relevant end-functionalization of the polymer. This commentary highlights the potentialities of coordinative chain transfer polymerization for this purpose. The concept of coordinative chain transfer polymerization is first introduced, followed by the use of aluminum and magnesium alkyls together with silane and borane as chain transfer agents for the synthesis of aluminum, magnesium, silicon and boron end-capped syndiotactic polystyrenes. The use of such end-capped macromolecules for additional functionalization is further discussed, with a focus on two topics: the formation of syndiotactic polystyrene functionalized with a variety of functions via organic transformation of the metal-carbon bond of the end-group and the formation of block copolymers with apolar and polar polymers.

In: Polystyrene: Properties, Performance... ISBN: 978-1-61209-117-4
Editor: James E. Gray

Chapter 1

POLYSTYRENE TRIBOLOGICAL PERFORMANCE: PROGRESS IN THE UNDERSTANDING OF POLYMERS ATTRITION DURING CHEMICAL ENGINEERING PROCESSES

Achraf Ghorbal [*] [a,g,f], ***Rim Ben Arfi*** [b,g], ***Férid Mokhtar*** [h], ***Adel Zrelli*** [a,c], ***Samir Ismaili*** [a,d], ***Federico Grisotto*** [e], ***Julienne Charlier*** [e] **and** ***Serge Palacin*** [e]

[a] Institut Supérieur des Sciences Appliquées et de Technologie de Gabès, Gabès, Tunisia

[b] Ecole Nationale d'Ingénieurs de Gabès, Gabès, Tunisia

[c] Unité de Recherche Environnement Catalyse et Analyse des Procédés, Ecole Nationale d'Ingénieurs de Gabès, Gabès, Tunisia

[d] Unité de Recherche de thermodynamique Appliquée, Ecole Nationale d'Ingénieurs de Gabès, Gabès, Tunisia

[e] Laboratoire de Chimie des Surfaces et des Interfaces, DSM/IRAMIS/SPCSI, Commissariat à l'Energie Atomique de Saclay, Gif-sur-Yvette, France

[f] Collaborateur au laboratoire de Chimie des Surfaces et des Interfaces, DSM/IRAMIS/SPCSI, Commissariat à l'Energie Atomique de Saclay, Gif-sur-Yvette, France

---

* Corresponding author: Dr. Achraf Ghorbal (e-mail: ghachraf@yahoo.com)

[g]Collaborateur à l'Unité de Recherches de Matériaux de Terres Rares, Centre National de Recherches en Sciences des Matériaux, Tunis, Tunisia
[h] Unité de Recherches de Matériaux de Terres Rares, Centre National de Recherches en Sciences des Matériaux, Tunis, Tunisia

Abstract

The issue of polymers films and particles friction, damage and attrition attracts both scientific and technological interest. In fact, polymers friction and attrition are of pivotal importance in many technological domains and applications such as powder technology, processes in fluidized beds, materials handling and transport processing (e.g., screw, pneumatic and hydraulic convey) and biopolymers and polymers processing. Moreover, polymer damage (e.g., attrition) has harmful effects on product quality, on environmental and health safety (*source of toxic dust*) and on the process reliability because of changes in particle properties such as particle shape and size distribution.

The present contribution will focus on the attrition analysis induced by the rub of hemispherical polystyrene particles against smooth and functionalized silicon walls. Several experimental and structural factors affecting attrition of polymeric particles such as the velocity, the applied normal force, polymer molecular weights, and the wall surface energy are discussed. A homemade device was used to carry out a single slide of the polystyrene hemispherical (*4 nm radius*) particle onto different silicon walls. Atomic force microscopy (AFM) was used so as to analyze the attrition appearance induced by the polymer damage. Nano-friction experiments were also achieved by AFM in lateral mode in order to distinguish the effect of interfacial interactions at the nanometric scale.

1. Introduction

Polystyrene (referred to as PS) is an engineering thermoplastic, used in many industrial applications [1]. In fact, polymeric materials like PS are used in range of mechanical components, e.g. cams, brakes and conveyors. Furthermore, such polymers are widely used in the field of chemical engineering applications, e.g. processes in fluidized beds [2, 3], powder technology [4], materials handling and transport processing [5-7], biopolymers and polymers processing [8] and so on. In many of these industrial processes PS is used as granulates and subjected to various stressing conditions. During last decades, the origin of polymeric particles damage phenomena has gradually been revealed. However, the whole description of chemical, physical, rheological and interfacial origins of aggregates damage is not yet entirely available [9-11]. Richardson et al. [12] showed that the force application method to the particles may affect the damage pattern. Moreover, Prasher [13] suggested that four basic particles damage ways may be identified, though it is sometimes difficult to identify the dominant mode in any given machine. The four basic particles damage ways are [12]:

- Impact: particle concussion by a single rigid force.
- Compression: particle disintegration by two rigid forces.
- Shear: produced by a fluid or by particle-particle interaction.
- Attrition: arising from particles scraping against one another or against a rigid wall.

Overall, attrition is an undesired phenomenon in industrial processes [14]. In fact, attrition has harmful effects on product quality and on the process reliability because of changes in particle properties such as particle shape and size distribution. Moreover, attrition is not only responsible of the product quality worsening but is sometimes also the source of harmful toxic dust [6].

Therefore, the issue of polymeric particles attrition attracts both fundamental research and technological interest. However, most of the published works kept huge importance to conditions under which damage by impact occurs [15-17]. In recent years, few articles have been published dealing with attrition under particle-particle or particle-wall friction effect [18-19] since it was established that during the flow of some particulate solids, the most extensive damage is caused by the shear deformation [20]. However, attrition by shear of most particle-particle or particle-wall systems is hard to be observed and to be quantified mainly because of particles dimensions, feeble attrition rate (e.g. undetectable at the macroscopic scale) and complex interfacial phenomena (e.g. surfaces energy and macromolecular chains influence on polymeric materials behavior) [21]. Furthermore, because of the important surface to volume ratio of particles, interfacial properties become increasingly important. In fact, the interface zone is expected to be the field of energy dissipation processes caused by the contribution of surface energy [22], and the mobility of macromolecules in the vicinity of the interface [23, 24].

Among the numerous methods commonly used for the surface energy change, three could be emphasized for their importance in fundamental research [25]:

- **Spin-coating method.** Spin-coating is the most widely used technique for organic coatings based on polymer solutions. The obtained thicknesses are generally ranged from 1 μm to 200 μm, but thinner films can be produced with dilute solutions and high spinning rates. The homogeneity of the spin-coated film strongly depends on the quality of the pristine solution and the chosen combination of solvent, spinning rate, and atmosphere. Spin-coating is ideal for temporary organic coatings, since the organic film is only physisorbed on the substrate.
- **Electrografting method.** Electrografting is a relatively new technique that allows the deposition of very thin organic films (typically ranged from one monolayer to 50 nm) via very strong substrate-molecule bridges [26, 27]. Often compared (or confused) with the anodic electro-deposition of conducting polymers from conjugated monomers, this grafting method is an electro-initiated process, which requires a charged electrode only for the grafting step, but not for thickening.
- **Self-assembly method.** Self-assembly is a widely used process for forming organic coatings [28]. Israelachvili [29] has deeply described the interactions of small molecular aggregates such as micelles, bilayers, vesicles and biological membranes, which form readily in aqueous solution by the spontaneous self-association or self-assembly of amphiphilic molecules. So as to observe spontaneous formation of true chemical linkages between molecules and solid substrates, the substrate is immerged in a solution containing the molecular precursors having a final head capable of specific interactions with surface. Thus, surfaces of oxides, noble metals, and silicon could be successfully functionalized using suitable reactive molecules, e.g. halogenosilanes, alkoxysilanes, organophosphonates, carboxylic acids, thiols, disulfides, selenols, alkenes, alkynes, etc. Self-assembly from solution or the gaseous state (for short molecules) provides very good coverage in short times. SAM's on conducting (noble metals) [30] and semi-conducting (silicon) [31, 32] surfaces constitutes doubtless the prototype of self-assembled layers because they are easy to realize.

In the next section, we describe the experimental methods, and then we focus on the description of the original device developed for attrition experiments and specimen's preparation. Chemical, physical and mechanical characterizations carried out at the macroscopic and nanometric scales will be also described.

2. Experimental Methods

2.1. Description of Attrition Device

A schematic diagram of a modified tribological device is shown in Figure 1. In design, the installation resembles an original pin-on-disc tribometer used by the master author (Ghorbal et al., 2006) in previous tribological studies. The main component of the attrition device is an alumina truck which is rolling on rails owing to frictionless bearings. For attrition experiments, a polymeric hemisphere is mounted on a stainless steel support. The tangential and the normal force are measured by means of orthogonal force sensors. Moreover, the displacement and velocity of the truck is controlled owing to a traction device.

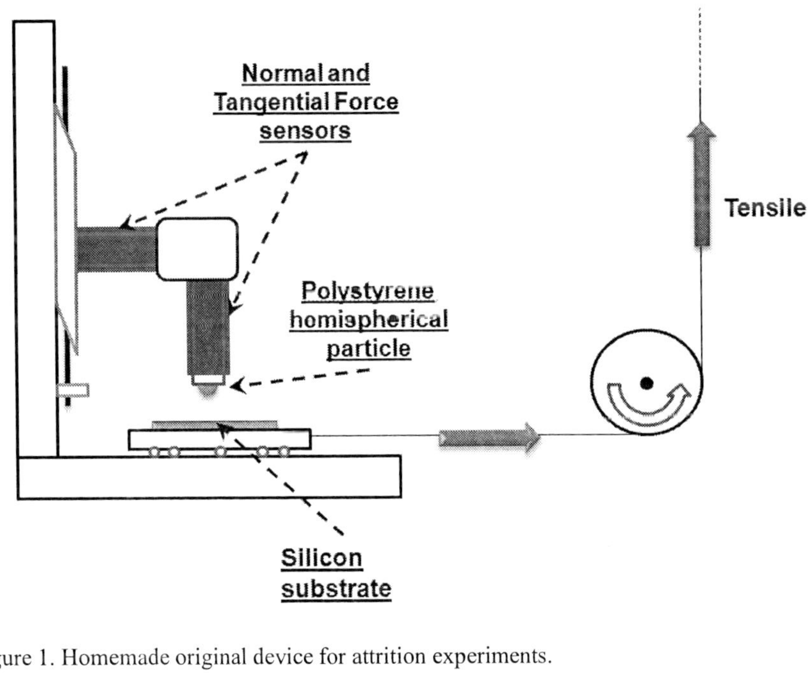

Figure 1. Homemade original device for attrition experiments.

In fact, this original attrition device has been essentially developed to carry out a single passage of the single hemispherical particle onto silicon substrate (simulating the wall). Due to the sphere on plan configuration, a constant contact area and consequently a defined contact pressure are maintained throughout the experiment. This device allowed measuring the tangential force (F_T) between the polymer particle and the wafer surface under controlled normal force (F_N) and sliding velocity (V). The linear sliding velocity was ranged from 0.1 mm/min to 25 mm/min (for instance, to simulate linear velocity of particles in contact with pipelines wall). The F_N applied on PS hemispheres varied from 1 ± 0.01 N to 7 ± 0.01 N (for example, to simulate particles density effect in conveying pipelines). All measurements were made at 22 ± 2 °C temperature and 35 ± 5 % relative humidity. At least five attrition measurements (one single rub for each measurement) were performed at any given set of experimental conditions to obtain the reported mean values.

2.2. Specimen Preparation

- Polymers and hemispheres preparation

Polystyrene was made by anionic polymerization. Styrene monomer (99%, Aldrich) was polymerized under a purified Argon atmosphere following the Lee et al. [33] method. The anionic polymerization apparatus was similar to the design of Ndoni et al. [34] employing a positive pressure of Argon gas, gastight fittings, and Teflon stopcocks to minimize the possibility of contamination by air while avoiding glass blowing and break seals. Purified cyclohexane was directly distilled into the reaction vessel under Argon atmosphere. The reactor of polymerization was immersed in a thermostated water bath so as to maintain the reaction temperature at 45 °C. The desired amount of the initiator, *sec*-butyllithium in hexane, was added by a gastight syringe to the cyclohexane solvent in the reaction vessel. The polymerization was initiated by adding purified styrene monomer. During the polymerization, the reaction mixture was transferred to Argon-filled Erlenmeyer flasks through a cannula for a programmed reaction time and terminated by injecting a small amount of degassed 2-propanol. The elapsed time between the introduction of styrene to the reactor and termination was recorded. Polystyrenes were precipitated in a 1/3 (v/v) mixture of 2-propanol/methanol.

Table 1 shows a summary of molecular weight averages (from GPC, *see 2.8 section*) of the polystyrene samples used in this study. Polystyrene sample

PS-385 showed the highest weight average molecular weight (384 985 g/mol). Sample PS-26 showed the lowest weight average molecular weight of 25 940 g/mol. [13] *C nuclear magnetic resonance* (AC Bruker NMR Instrument) studies revealed that all the polystyrene samples used were atactic and linear without any branching or crosslinked sites.

Table 1. Average molecular weights derived from gel-permeation chromatography (all molecular weights are ×1000 g/mol)

PS name	M_w	M_n	M_w/M_n
PS-26	25940	23478	1.11
PS-118	118098	106882	1.10
PS-245	245326	219016	1.12
PS-385	384985	354690	1.09

PS hemispheres were made by a molding process. PS hemispheres (4 mm radius) were obtained by molding the polymer (obtained by anionic polymerization) at 220 °C during 20 minutes. Before attrition experiments, PS hemispheres were conditioned at 22 ± 2 °C temperature and 35 ± 5% relative humidity during at least 24 hours.

- Surfaces preparation

Substrates, for macroscopic and nanometric experiments, were (2 × 3) cm^2 slices of SiO_2/Si(100) one-side-polished silicon wafers (supplied by ACM, France). For the results relevance, we used silicon slices cut from the same wafers compartment.

- Substrates preparation for attrition experiments : **piranha treatment** and **self assembly** method

For attrition experiments, it is essential to have extremely flat and well characterized surfaces of walls, i.e., substrates with the same surface energy and roughness. In fact, we want to detect the beginning of polymer transfer onto hydrophilic and hydrophobic surfaces at the nanometer scale, and therefore the roughness of our substrates has to be as small as possible. These conditions are achieved, with silanol-enriched silicon wafers (SiO_x-OH) and methyl-terminated silicon surfaces (-CH_3), which are here obtained after the piranha treatment and the adsorption of SAM's following the Haidara et al [35] method, respectively.

In fact, slices of silicon substrates are cleaned with absolute ethanol and immersed for 30 min in warm (60 °C) piranha solution (3:7 v/v mixture of 30 % H_2O_2 and H_2SO_4 in volume) to hydroxylate the surface. The sulfuric acid and the hydrogen peroxide are powerful oxidizing agents that react rapidly with almost any kind of organic compound. The hydrogen gas is one of the products of this type of reaction. Immediately after the immersion of silicon substrates into piranha solution, visible gaseous bubbles evolved from the silicon slices, indicating the presence of organic contaminants on the surface [36]. After piranha treatment, further hydroxylation is achieved by a thorough cleaning and storage of the plates for about 24 h in deionized and twice distilled water (*Caution: The piranha solution is a strong oxidant and reacts violently with organic substances, it should be handled with extreme care!*) [37]. The substrates are then dried under nitrogen stream before immediate use for experiments.

However, hydrophobic (methyl-terminated) substrates are obtained by immersion of activated surfaces (hydroxylated) into the dodecyltrichlorosilane (*DTS*, $C_{12}H_{25}SiCl_3$) solution for an overnight (about 12 h) adsorption in carbon tetrachloride. The piranha pretreatment is aimed to develop a high silanol density at the silicon surface where the chlorosilane will self-assemble upon hydrolysis [38]. After that adsorption step, the substrates are sonificated in carbon tetrachloride, rinsed in carbon tetrachloride solution to remove any excess of physisorbed silanes, rinsed with deionized water, and then dried under a nitrogen flow before immediate use for experiments.

– Substrates preparation for nanometric characterizations : **spin-coating method**

Solutions of PS/toluene were prepared by weight in concentrations of 10.0 wt.%. The samples were spun in open air for 60 s at 1000 rpm speed onto cleaned silicon (100) wafers. As spin-coating progressed and the liquid layers thicknesses became of order 1 µm or less, a progression of time-varying interference colors was seen.

Since the color changes stopped within approximately 20 s of spin-up for all concentrations, it was inferred that the spinning time of 60 s was adequate for providing the minimum final film thickness. Films were allowed to dry in air, and then placed in a vacuum oven at 140 °C (above PS glass transition temperature) for 1 h, for removing residual toluene [39].

2.3. Contact Angles Measurements and Surface Free Energy

An automatized Krüss G2 goniometer employing the sessile drop technique was used for equilibrium contact angles, θ, measurements. This instrument is equipped with automatic image acquisition and contact angle computation software. Water, formamide and α-bromonaphtalene were used as liquids to obtain corresponding contact angles values on both sides of the drop image. This process was repeated five times. Measurements were carried out in open air at 22 ± 2 °C temperature and 35 ± 5 % relative humidity. Droplets volume was equal to 10 μl.

The equilibrium contact angle, θ, for a liquid drop on an ideal, homogeneous, planar, and rigid surface is related to the various interfacial tensions by the Young's equation:

$$\gamma_{lv}\cos\theta = \gamma_{sv} - \gamma_{sl} \qquad \text{Eq. (1)}$$

Where γ_{lv} is the surface tension of the liquid in equilibrium with its saturated vapor, γ_{sv} surface tension of the solid in equilibrium with the saturated vapor of the liquid, and γ_{sl} is the interfacial tension between the solid and the liquid.

The Fowkes [40-43] method improved by Owens and Wendt [44] based on the contact angles of a series of test liquids can be combined with Young's equation [Eq. (1)] to give

$$\gamma_l(1+\cos\theta) = 2(\gamma_s^d\gamma_l^d)^{\frac{1}{2}} + 2(\gamma_s^p\gamma_l^p)^{\frac{1}{2}} \qquad \text{Eq. (2)}$$

where $\gamma = \gamma^d + \gamma^p$, the subscript *(l)* refers to a series of test liquids with known dispersive and polar components while the subscript *(s)* refers to solids with their surface free energy dispersive γ^d_s and polar γ^P_s components.

Test liquids surface tensions and their dispersive and polar components [45, 47] are given in Table 2.

Table 2. Surface tensions and components (mJ/m²)

Liquids	Surface tension	Dispersive component	Polar component
water	72.8	21.8	51
Formamide	58	39	19
α-bromonaphtalene	44.4	44.4	0

2.4. Atomic Force Microscopy Examinations

- Topographical measurements of attrition

Topography measurements of transferred (under attrition effect) PS onto silicon surfaces (simulating the wall) were realized using tapping mode atomic force microscopy referred to as TMAFM [48, 49]. This dynamic force microscopy mode allowed scanning surface with constant oscillating amplitude introduced in the vertical direction with a feedback loop keeping the average tapping normal force constant [50]. Phase and height images were recorded simultaneously using a commercial Dimension 3100 AFM equipped with a Nanoscope IIIa controller (Digital Instruments, Santa Barbara, CA). A square pyramidal silicon nitride tips (Veeco probes) were mounted on 145 μm long single beam cantilever with a resonance frequency of approximately 150 kHz and a spring constant of about 5 $N.m^{-1}$. The scan rate was in the range 0.5–1 Hz with a scanning density of 512 lines/frame.

- Contact mode measurements

Experimental Details. For nano-adhesion measurements a commercial atomic force microscope (D3100, Nanoscope IIIa controller, Digital Instruments, Santa Barbara, CA) was used. AFM experiments were performed at a relative humidity controlled within the range of 35-40 % in order to keep the same capillary force contribution.

Nano-adhesion Measurements. Adhesive force measurements are performed in the so-called force calibration mode. In this mode, "approach" or "force distance" curves are obtained by recording the cantilever deflection, for an example, see Figure 2.

The deflection of the cantilever (*i.e.,* the cantilever bends vertically toward or away from the sample in response to attractive and/or repulsive forces acting on the tip.) is detected by an optical device (four quadrants of photodiodes) while the tip is vertically moved forward and backward thanks to a piezoelectric ceramic (or actuator). Thus, one can obtain a deflection distance curve [51].

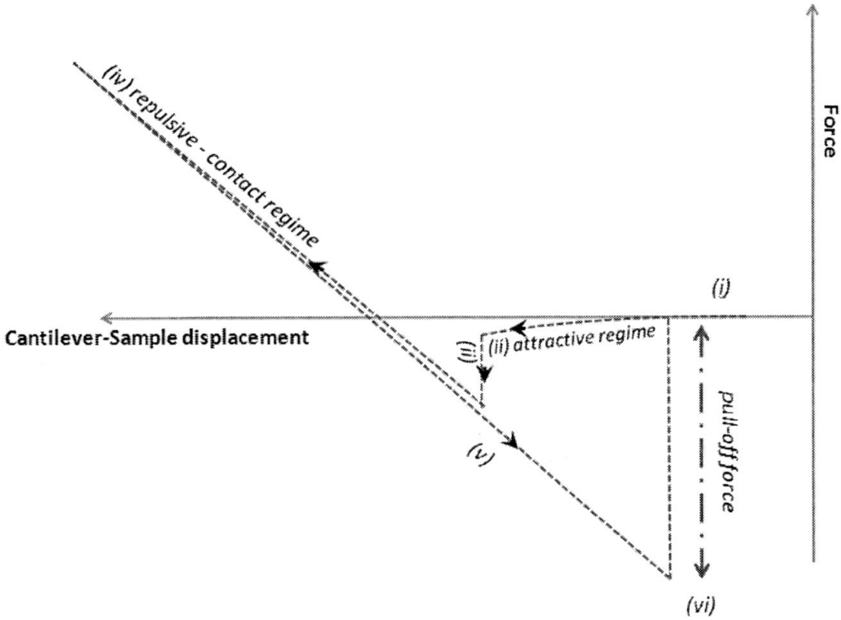

Figure 2. An “approach curve” or “force distance” curve displays the vertical cantilever bending versus lever-sample displacement.

Approach curve [52]. The horizontal axis gives the distance the piezoelectric (and hence the sample) travels, and the vertical axis gives the tip deflection.

i. The cantilever and sample are initially far apart and no forces act: as the piezoelectric extends, it approaches the tip, which is at this point in free air and hence shows no deflection. This is indicated by the flat portion of the curve.
ii. As the lever is brought close to the sample, the tip senses attractive forces which cause the end of the cantilever to bend downward, thus signifying a negative (attractive) force.
iii. The attractive force gradient exceeds the spring constant of the cantilever at this point, and this instability causes the tip to snap into contact with the sample.
iv. The lever-sample displacement can continue to be reduced. Since this tip is in repulsive contact with the sample, the front end of the cantilever is pushed further and further upward. The force

corresponds to the externally applied load: this is represented by the sloped portion of the curve.

v. The motion is reversed. Adhesion between the tip and sample maintains the contact although there is now a negative load.

vi. Finally the tensile load overcomes the adhesion or pull-off force and the tip snaps out of contact with the sample.

The adhesive force was determined from the difference in the vertical deflections of the piezoelectric tube, Δz, between the points where hard-wall contact was achieved and where the tip snapped free from the surface. The raw voltage data is converted to normal forces (F_N) with Hooke's law for small cantilever deflections [50, 53-55]:

$$F_N = K_n . \Delta z = K_n \frac{\Delta V}{S_n} \qquad \text{Eq. (3)}$$

where ΔV is the photodiode voltage response, k_n is the normal spring constant, and S_n is the normal sensitivity, or the ratio of the photodiode vertical response to the displacement of the piezoelectric tube. To quantify the adhesive forces, approximately 50 force-displacement curves were obtained for each interaction from a total of five contact points in the scan area.

A rigid reference sample (cleaned silicon substrate) is used to scale the approach curve in deflection by fixing to unity the slope value of the contact line.

- Friction mode measurements

Experimental Details. For nano-friction measurements a commercial atomic force microscope (D3100, Nanoscope IIIa controller, Digital Instruments, Santa Barbara, CA) was used. AFM experiments were performed at a relative humidity controlled within the range of 35-40 % in order to keep the same capillary force contribution.

Nano-friction Measurements. For measurement of the friction force being applied at the tip surface during sliding, the left hand and right hand sets of quadrants of the photodetector are used. In the so-called friction mode, the sample is scanned back and forth in a direction orthogonal to the long axis of the cantilever beam. A friction force between the sample and the tip will produce a twisting of the cantilever. As a result, the laser beam will be reflected out of the plane defined by the incident beam and the beam reflected vertically from an untwisted cantilever. This produces an intensity difference of the laser beam received in the left hand and right hand sets of quadrants of the photodetector. The intensity difference between the two sets of detectors is directly related to the degree of twisting and hence to the magnitude of the friction force [50].

Friction forces mean values were extracted from lateral diffraction (friction loops: the trace minus retrace value (TMR)) over scan areas of (1 μm x 1 μm).

Friction forces were averaged for more than 50 data points per sample at different locations on the surface [56].

- AFM calibration

Cantilevers calibration. The systematic calibration procedure of the AFM measurements was performed to produce quantitative data. In literature, numerous papers have described destructive and non-destructive methods to determine the spring constant value of rectangular and V-shaped cantilevers [57-60]. In our case, we focused on two non-destructive methods, which have been rigorously studied and compared: the first one is related to the use of rectangular reference cantilevers [61] and the second is the unloaded resonance method of Sader et al. [62], referred to as the normal Sader method [63]. Both calibration methods have been used and compared to calibrate the triangular-shaped cantilevers (supplied by Nanosensor, Germany) used in this study. The effective spring constant of cantilevers was about 0.21 $N.m^{-1}$, whereas the value specified by the supplier was equal to 0.58 $N.m^{-1}$.

Tip radius of curvature determination. We deduced the radius on the basis of the shape revealed on a scanning electron microscopc picture taken at the end of the measurements.

Our tip radius was estimated to be 45 ± 5 nm. It is important to note that the adhesion force has been checked regularly on a reference silicon surface so as to reveal any contamination of the tip or eventually a change of the tip radius during experiments.

2.5. Infrared Spectroscopy Measurements

The infrared spectrum of the isotropic state of polystyrene (PS bulk) was obtained by analyzing PS bulk samples in KBr pastilles in the simple Transmission device (Brüker IFS 48 spectrometer). Scans number was equal to 200 with a 2 cm^{-1} resolution. Spectra were recorded with a mercury-cadmium-telluride (MCT) detector.

2.6. Differential Scanning Calorimetry Measurements

The glass transition temperature (*T*g) of each PS sample was measured in the range 0 °C to 170 °C using a differential scanning calorimeter (DSC) (model Mettler TA 4000). Each PS sample was preheated to 170 °C and maintained at that temperature for 10 min in order to remove any thermal history, then cooled at 10.0 $°C.min^{-1}$ to 0 °C, and once more heated at 10.0 $°C.min^{-1}$ ramping rate to 170 °C. During the measurements, dry nitrogen gas was used for purging. In each run, a sample of about 10 mg was used.

2.7. Rheological Measurements

The rheological behavior of polystyrenes has been studied using the parallel plate Physica MCR500 rheometer (Anton Paar Gmbh). Rheological measurements were performed on different PS melt at 220 °C. Shear rate values were ranging from 0.01 s^{-1} to 10 s^{-1}. The parallel plate diameter and the gap were respectively about 25 mm and 1 mm.

2.8. Gel-Permeation Chromatography (GPC) Measurements

Accurate determination of the molecular weight of a polymer is of considerable importance on polymers viscoelastic behavior. Gel-permeation chromatography was used in this study to accurately determine the relative molecular weights and molecular mass distribution of the polystyrene samples. Measurements were made using a Waters Associates (Model 510) Gel-Permeation Chromatograph operating at 30 °C. Three columns were used in series packed with styragel beads. The solvent used was tetrahydrofuran (THF). The M_w and M_w/M_n values determined by GPC are listed in Table 1. The PS molecular weights were calculated by calibration with eight different PS standards with molecular weights between 1200 $g.mol^{-1}$ and 970 000 $g.mol^{-1}$.

3. Results and Discussion

The interface zone is expected to be the field of energy dissipation processes [64, 65] caused by the contribution of surface energy [66] and opposing surfaces macromolecular interactions, and the mobility of macromolecules in the vicinity of the interface [67]. Moreover, it has been demonstrated that the terminal chain group chemistry of rubbed surfaces [68] and polymers chains mobility [69, 70] can have a significant impact on the friction [71].

Furthermore, investigations of such systems are complicated by the molecular dimensions of the transfer films (confinement). In addition, it has been established that the rheology of molecular films is usually completely different from bulk [72, 73] and that their properties cannot necessarily be known by an extrapolation of polymer bulk [74–77].

The objectives of the first part of the current section are to specify the role of interface in friction and attrition appearance of amorphous polymer (polystyrene). By reducing the surface roughness, the role of chemical interactions between opposing surfaces is magnified and it becomes possible to analyze the effect of interfacial interaction on attrition [78]. The role interface chemistry is then investigated by comparing friction and damage appearance of polystyrene against hydrophobic walls (methyl-terminated) and hydrophilic walls (hydroxyl-terminated).

Wettability measurements were achieved to determine surface energy components of piranha treated and grafted (by self-assembly method) substrates and the 4 PS. Surface energy components of used materials are gathered in table 3. From table 3 we can immediately see that PS-26, PS-118, PS-245 and PS-385 surface energies are identical.

These surfaces energies are in agreement with reference data reported for such surfaces [79, 80]. Moreover, Infrared spectroscopy measurements confirm that 4 PS have an identical infrared spectrum and thus they have the same chemical composition.

Figure 3 illustrates the transmission spectrum of polystyrene bulk corresponding to the isotropic state of the polymer. Detailed information about infrared bands attribution was obtained from literature [81, 82]. However, assignments of peaks around 1452, 1493 and 1601 cm^{-1} were ambiguous. Therefore, we presented here only pertinent wavenumbers, particularly: CH_2 asymmetric stretching $\nu_{as}(CH_2)$ around 2923 cm^{-1}, CH_2 symmetric stretching

$\nu_s(CH_2)$ around 2849 cm^{-1}, aromatic CH stretching $\nu(CH)_{ar}$ around 3026, 3060 and 3082 cm^{-1}. [21]

Table 3. Dispersive and polar components and surface free energy of materials used in attrition experiment ($mJ.m^{-2}$)

Sample name	Surface energy ($\pm$ 2 $mJ.m^{-2}$)	Polar component ($\pm$ 1 $mJ.m^{-2}$)	Dispersive component ($\pm$ 1 $mJ.m^{-2}$)
PS-26	41	1	42
PS-118	41	1	42
PS-245	41	1	42
PS-385	41	1	42
Piranha treated wall	79	43	36
DTS grafted wall	20	1	21

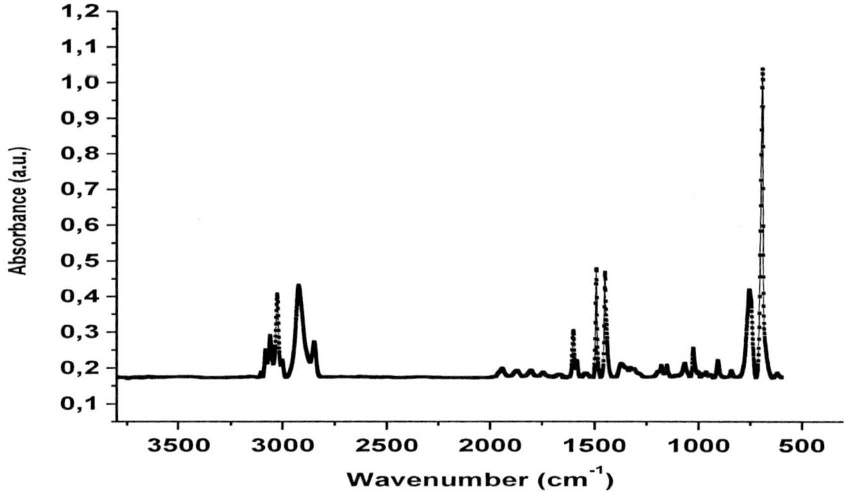

Figure 3. Infrared Transmission spectrum of polystyrene bulk.

3.1. Discussion of Rubbing Results

Wettability and infrared spectroscopy measurements suggest that 4 PS hemispheres have the same chemical affinity to each silicon wall. Hence, the main advantage of the investigated system is to have comparable interfacial interactions between PS-26, PS-118, PS-245 or PS-385 and each silicon wall.

Table 4 and table 5 report the friction coefficient values (μ: tangential force divided by normal force) of 4 PS in contact with the hydroxyl-terminated and the methyl-terminated silicon walls, respectively.

Table 4. Friction coefficient μ values for 4 PS in contact with the hydrophilic wall, for different normal forces and friction speeds ($\Delta\mu = \pm$ 0.02)

V ($mm.min^{1}$)	0.1				1				10				25			
F_n (N)	1	3	5	7	1	3	5	7	1	3	5	7	1	3	5	7
PS name																
PS-26	0.38	0.40	0.42	0.43	0.40	0.43	0.47	0.46	0.42	0.45	0.46	0.49	0.52	0.53	0.55	0.55
PS-118	0.31	0.34	0.36	0.36	0.33	0.35	0.36	0.38	0.32	0.35	0.38	0.39	0.34	0.34	0.38	0.37
PS-245	0.27	0.29	0.32	0.33	0.29	0.29	0.33	0.33	0.30	0.31	0.32	0.34	0.31	0.32	0.34	0.34
PS-385	0.24	0.24	0.26	0.26	0.24	0.26	0.27	0.29	0.26	0.27	0.29	0.30	0.26	0.26	0.30	0.31

Table 5. Friction coefficient μ values for 4 PS in contact with the hydrophobic wall, for different normal forces and friction speeds ($\Delta\mu = \pm$ 0.02)

V ($mm.min^{-1}$)	0.1				1				10				25			
F_n (N)	1	3	5	7	1	3	5	7	1	3	5	7	1	3	5	7
PS name																
PS-26	0.11	0.13	0.13	0.14	0.13	0.14	0.16	0.17	0.14	0.16	0.17	0.17	0.15	0.17	0.19	0.20
PS-118	0.10	0.13	0.12	0.13	0.12	0.13	0.15	0.15	0.13	0.15	0.16	0.17	0.15	0.16	0.17	0.18
PS-245	0.09	0.12	0.13	0.13	0.12	0.12	0.13	0.14	0.13	0.14	0.13	0.16	0.14	0.14	0.16	0.18
PS-385	0.10	0.11	0.12	0.13	0.11	0.12	0.13	0.13	0.13	0.13	0.14	0.15	0.14	0.15	0.15	0.16

Influence of the molecular weight on friction. Taking into account the experimental error, quite identical friction coefficients are obtained for 4 PS whatever the speed and normal force onto hydrophobic walls. However, onto hydrophilic walls, PS-26 showed higher friction coefficient values comparing to PS-118, PS-245 and PS-385 ones. Thus, it is reasonable to assume that the influence of chains length on friction coefficient is negligible on hydrophobic surfaces and quite important for low molecular weights on hydrophilic walls. Such phenomenon could be dependent on the molecular weight of the critical entanglement, M_c, of PS $\approx 31.2\ 10^3$ g.mol^{-1}. [83] In fact, short macromolecular chains could show higher friction coefficients under confinement and chains orientation effect. Thus, under such conditions polymers behave like a solid due to high inter-chains friction phenomenon. Consequently, rubbing is less easy and the friction coefficient is then higher.

Influence of the applied normal force on friction. As seen from table 4 and table 5, whatever the sliding speed and the terminal surface chemistry, the applied load effect is insignificant for 4 polystyrenes.

Influence of the velocity on friction. For rubbed polystyrenes against hydrophilic and hydrophobic surfaces, the friction coefficients increased with the sliding velocities ranging from 0.1 to 25 $mm.min^{-1}$. For example for the 3 N applied normal force, the PS-26 friction coefficient increased from 0.4 at V = 0.1 $mm.min^{-1}$ to 0.54 at V = 25 $mm.min^{-1}$, on hydrophilic surfaces. On hydrophobic surfaces and for the same applied normal force, the PS A friction coefficient varied from 0.13 at V = 0.1 $mm.min^{-1}$ to 0.17 at V = 25 $mm.min^{-1}$. These experimental results are inconsistent with the basic third classical law of friction, because the kinetic friction of polystyrenes is dependent of the sliding velocity. However these results are in agreement with previous studies which have shown the friction dependence of sliding speed for example at clean surfaces or atomic level [84-86]. This speed dependence of the friction coefficient can be explained by the intrinsic properties of polystyrenes which are viscoelastic materials. In fact the amount of energy loss depends markedly on the velocity at which deformation takes place [87, 88].

Nevertheless, the change of the nature of interfacial interactions by using wafer surfaces terminated by -OH and $-CH_3$ end groups affected profoundly the friction. It should be pointed out that the variation of friction coefficient on both surfaces under the same experimental conditions is comparable. However, hydrophilic surfaces friction coefficient values are always larger than those obtained with hydrophobic surfaces. These results are consistent with our expectation that the interaction between polystyrenes and hydrophilic groups will be stronger than that between polystyrene and hydrophobic groups, whereas the interaction should be weakest.

We emphasize that, with the hydrophobic surfaces, the friction coefficients values are always lower than those obtained with the hydrophilic surfaces. Thus, in the present study, wettability and friction results suggests that the chemical nature at the top of the wall surface influence the adhesion and the interfacial friction. What about the attrition appearance? Is it influenced by the interfacial interactions?

However, it is of prime importance to observe PS's behavior change during the rub on walls, but suggestions based on such observations aren't enough to definitely prove the attrition happening. Hence, in the following parts of this chapter, we will focus on the attrition appearance prove and interpretation of process governing this phenomenon.

3.2. Discussion of Attrition Results

We have to remember that attrition experiments were performed to evaluate the beginning of polymeric particles damage against a model wall (simulating a pipeline wall). In this attempt, we varied, at first, the applied normal force, hemispheres velocity and the wall surface energy so as to simulate, for instance, the effect of particles density, the velocity in conveying pipelines and different surface treatments, respectively.

With regard to the morphological analysis of macromolecular film transfer (attrition appearance), the atomic force microscope (AFM) has played a very important role to image the transfer layer. During the last decade, the AFM was a very useful device when the transfer observation was no longer possible by optical microscope.

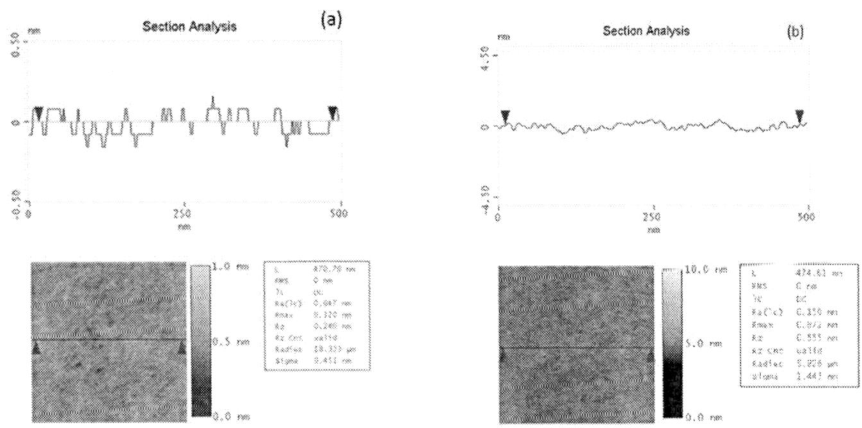

Figure 4. Tapping mode AFM (a) height image section analysis of hydrophilic wall (b) height image section analysis of hydrophobic wall. The scan scale is 500nm.

Using AFM, images were taken in tapping mode. The morphology of the piranha treated and methyl grafted surfaces before sliding against polystyrenes observed with AFM are shown in figure 4. The height (see Figure 4a) image of the piranha treated surface shows a very homogeneous and clean silicon wall. Figure 4b, additionally indicates that no aggregates were formed during the grafting of dodecyltrichlorosilane molecules. In addition, the surface average roughness (Ra) was calculated to be about 0.05 and 0.15 nm in a (500 nm x 500 nm) scan area, respectively, for the hydroxyl-terminated and methyl-terminated walls, which shows a highly ordered SAM formed onto the surface. These results confirm that hydrophilic and hydrophobic walls were homogeneous and especially little rough.

- Attrition on hydrophilic walls

Tribological studies of polymers have shown that under sufficiently contact pressure the polymer surface becomes damaged and the transfer appears onto the opposing surface [89-91]. Table 6 report necessary critical conditions, revealed by AFM characterization, for attrition appearance onto the hydroxyl-terminated silicon walls. On the basis of AFM topographical observations and reported results in table 6, one can confirm that no attrition appearance (for 4 PS) has been detected for normal force inferior or equal to 3 N and 1 N and for sliding speeds of about 0.1 $mm.min^{-1}$ and 1 $mm.min^{-1}$, respectively.

Attrition observations under 0.1 $mm.min^{-1}$ sliding speed. Onto hydrophilic surfaces, the PS-25 attrition has been observed for the first time under 5N normal force and 0.1 $mm.min^{-1}$ sliding speed. Moreover, PS-118 and PS-245 attrition has been detected for the first time under 7N normal force (higher normal force comparing to PS-26) and 0.1 $mm.min^{-1}$ sliding speed, whereas no damage of the PS-385 has been observed under the same experimental conditions.

Table 6. Critical conditions (colored cells) for attrition appearance onto hydrophilic wall (as function of normal load), for 4 PS ($\Delta\mu = \pm 0.02$)

V ($mm.min^{-1}$)	0.1				1				10				25			
F_n (N)	1	3	5	7	1	3	5	7	1	3	5	7	1	3	5	7
PS name																
PS-26	0.38	0.40	0.42	0.43	0.40	0.43	0.47	0.46	0.42	0.45	0.46	0.49	0.52	0.53	0.55	0.55
PS-118	0.31	0.34	0.36	0.36	0.33	0.35	0.36	0.38	0.32	0.35	0.38	0.39	0.34	0.34	0.38	0.37
PS-245	0.27	0.29	0.32	0.33	0.29	0.29	0.33	0.33	0.30	0.31	0.32	0.34	0.31	0.32	0.34	0.34
PS-385	0.24	0.24	0.26	0.26	0.24	0.26	0.27	0.29	0.26	0.27	0.29	0.30	0.26	0.26	0.30	0.31

Attrition observations under 1 $mm.min^{-1}$ sliding speed. When the hemispheres velocity is increased to 1 $mm.min^{-1}$, the PS-26 attrition has been observed under 3N normal force. Furthermore, PS-118 and PS-245 attrition appearance has been observed under 5N normal force. It is important to note, that the PS-385 attrition has been observed for the first time under 7N normal force and 1 mm.min^{-1} sliding speed.

Apparent contact area and contact pressure. The apparent contact area and corresponding mean contact pressure were determined in terms of the hemispheres radius, applied normal force, and elastic properties of surfaces using the classical Hertz (1882) theory. In fact, if a hemisphere with an applied normal force F_N is in contact with a plane surface, the elastic material (in our case the PS hemisphere) will deform to make contact over a circle of the radius (a) given by Hertz (1882) equation:

$$a = \left(\frac{3.F_N.R}{4.E}\right)^{\frac{1}{3}} \quad \text{Eq. (4)}$$

Where R is the radius of curvature of particles and E is the effective modulus, which is given by:

$$\frac{1}{E} = \left(\frac{1-\nu_1^2}{E_1} + \frac{1-\nu_2^2}{E_2}\right) \quad \text{Eq. (5)}$$

Here, suffices 1 and 2 refer to the PS hemisphere and to the silicon wall, respectively.

PS's elastic modulus (E_1) was assumed to be equal to 3.2 GPa, and that of Silicon wall (E_2) equal to 130 GPa. Poisson's ratios of 0.40 and 0.25 were assumed for polymeric particles and silicon wall, respectively. Table 7 depicts the circle radius given by Hertz equation, the contact area and the corresponding normal stress, for the different forces. Assuming that 4 PS have the same mechanical properties (in particular Young's modulus), contact area is expected to be the same under similar experimental conditions.

In fact, one has to note that during attrition experiments, the normal stress has never been superior to the yield point. Under such conditions, the PS deformation could be then considered as purely elastic. However, rubbing with normal stresses values inferior to the 95 MPa yield stress value [92] does not prevent plastic deformation of PS surface asperities. As reported in previous publications [93], this behavior could occur under the temperature increase effect at the top of polymeric surfaces.

Table 7. Contact areas and corresponding contact pressures under the different applied normal forces during attrition experiments.

F_N (N)	a (mm)	Area (mm^2)	Normal stress (MPa)
1	0.0932	0.0273	36.66
3	0.1344	0.0567	52.87
5	0.1593	0.0798	62.68
7	0.1783	0.0998	70.12

Applied Normal Force Effect

In this paragraph, we will focus results analysis of the normal load effect on a 1 $mm.min^{-1}$ sliding velocity value. TMAFM analysis, on wafers rubbed at 1N normal force, showed nearly no change of the silicon wall aspect and roughness compared to the piranha treated wall, whatever the PS molecular weight and the sliding velocity value. Thus, it seems evident that 36.66 MPa normal stress value is not enough to induce PS hemispheres surface damage, and therefore attrition appearance.

However, TMAFM scans of walls rubbed under 3N normal force show completely changed pictures. Figure 5 shows PS-26 transferred onto the hydrophilic silicon wall under 3 N normal load and 1 $mm.min^{-1}$ sliding velocity. Observations based on TMAFM images and table 6 confirm then that PS attrition appearance (revealed by material transfer onto the wall) is possible only for normal forces values equal or superior to 3N. We should remember that the main mechanisms of transfer of thermoplastic polymers like PS are pull-out, disentanglement and reptation of macromolecular chains [8, 94].

Moreover, we should highlight the fact that attrition appearance of PS-26, PS-118, PS-245 and PS-385 occurred for 1.29 N, 1.8 N, 1.65 N and 2.03 N tangential forces values, respectively. On the basis of these values, one can say that quite identical shear stress (a mean value of about 21 MPa, *taking into account experimental uncertainties*) was needed to generate attrition of 4 PS.

Thus, it is well established that a critical shear stress is needed in order to disentangle, pull-out, reptate and probably break PS's macromolecular chains and therefore generate attrition.

It has been shown that a critical interfacial shear is needed for the disentanglement of polystyrene chains and therefore implies the transfer appearance onto wafer surface. In addition, this critical interfacial shear may well induce an interfacial temperature increase and as a consequence soften the polymer surface which allows the transfer on opposing surface. In fact, a sufficient increase of temperature will decrease the polymer surface cohesion and consequently facilitate the reptation and the disentanglement of polymer chains. As a result, the transfer occurs when the polymer crossover from a solidlike to a liquidlike material showing a flow behavior [95]. Indeed, these results show that the appearance of PS's attrition is directly dependant on the applied normal force.

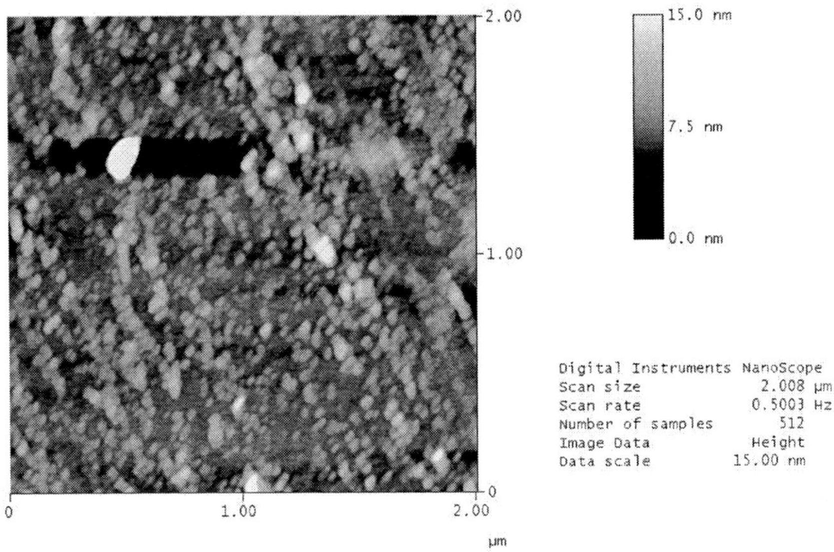

Figure 5. AFM tapping mode images of PS-26 attrition appearance on hydrophilic silicon wall under critical interfacial shear. Image area 2µm x 2µm. Height range is 15nm. The transfer appears under an applied normal force equal to 3 N and for a sliding velocity of about 1 $mm.min^{-1}$.

Hemispheres Velocity Effect

In this paragraph, we will focus results analysis of the sliding speed velocity effect on a 5 N normal load value. On the basis of results reported in table 8, it clearly appears that attrition occurs for 4 PS under 5 N normal force. However, changes in the silicon wall topography after rubbing under 5 N were revealed for PS-26, PS-118, PS-245 and PS-385 at 0.1 $mm.min^{-1}$, 1 $mm.min^{-1}$, 1 $mm.min^{-1}$ and 25 $mm.min^{-1}$, respectively.

Then, it may be argued that a critical velocity is needed so as to generate attrition. In fact, the velocity increase may induce a temperature increase at the hemisphere/wall interface, the softening of polymer surface (decrease of the surface cohesion) and therefore the attrition appearance.

Table 8. Critical conditions (colored cells) for attrition appearance onto hydrophilic wall (as function of sliding velocity), for 4 PS ($\Delta\mu = \pm$ 0.02)

F_n (N)	1				3				5				7			
V (mm.min^{-1})	0,1	1	10	25	0,1	1	10	25	0,1	1	10	25	0,1	1	10	25
PS name																
PS-26	0.38	0.40	0.42	0.52	0.40	0.43	0.45	0.53	0.42	0.47	0.46	0.55	0.43	0.46	0.49	0.55
PS-118	0.31	0.33	0.32	0.34	0.34	0.35	0.35	0.34	0.36	0.36	0.38	0.38	0.36	0.38	0.39	0.37
PS-245	0.27	0.29	0.30	0.31	0.29	0.29	0.31	0.32	0.32	0.33	0.32	0.34	0.33	0.33	0.34	0.34
PS-385	0.24	0.24	0.26	0.26	0.24	0.26	0.27	0.26	0.26	0.27	0.29	0.30	0.26	0.29	0.30	0.31

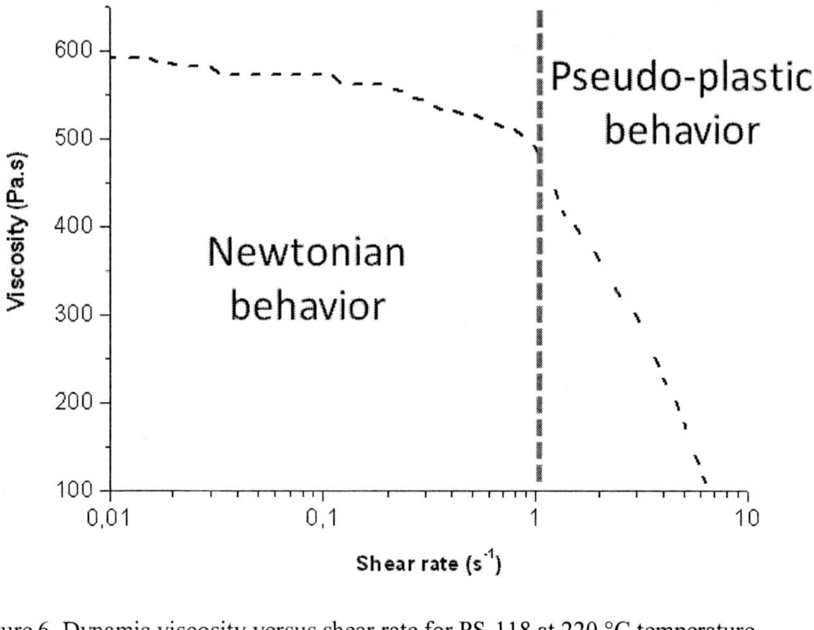

Figure 6. Dynamic viscosity versus shear rate for PS-118 at 220 °C temperature.

Moreover, under such conditions, PS's surface in contact with the wall could behave like a liquid and consequently the rubbing speed may affect the rheological behavior of polymeric hemispheres. In fact, rheological characterization showed that regardless the molecular weight, PS has a pseudo-plastic behavior for shear rate values superior to 0.9s^{-1} (see PS-118 example depicted in Figure 6). Shear rate value of about 1.1 s^{-1} corresponds to 66 mm.min^{-1} linear speed value, taking into account a 1mm rheometer gap. This observation clearly shows that an attrition appearance under a shear-thinning effect of PS could be then a plausible hypothesis.

Furthermore, more complex relationships between applied normal force, rubbing velocity and surface damage are most often observed. For example, Ashby et al. (1990; 1991) showed that, during friction, the local temperature at the interface is substantially higher than that of the environment, and is also enhanced at the asperity contacts by transient flashes. In fact, Ashby's (1990; 1991) empirical equations show that the temperature increase at the interface is proportional to the applied normal force and the sliding speed. In fact, Ashby showed that the heat generated *Q* (***per unit*** *time and* ***per unit area***) during friction is given by:

$$Q = \mu . F . V . (A_n)^{-1} \qquad \text{Eq. (6)}$$

where μ is the friction coefficient, *F* the applied normal force, *V* the sliding velocity and An the unit area.

Such results confirm our correlations und clearly show that the temperature increase at the interface is a key parameter in the attrition process.

Polymer Molecular Weight Effect

TMAFM results clearly indicated the significance of PS hemispheres molecular weights on the attrition appearance. In fact, we observed that attrition occurs more easily for low molecular weights, thus, one can suggest that such behavior is maybe related to polymers macromolecular chains mobility. It is well known that viscoelastic properties of polymers depend on the molecular weight. For example, according to Fox-Flory (1948; 1951) relationship, glass transition temperature (T_g) is related to macromolecular chains length (M_n) by

$$T_g = T_{g\infty} - \frac{K}{M_n} \qquad \text{Eq. (7)}$$

where $T_{g\infty}$ is the glass transition temperature for an ideal chain for infinite length and K is a characteristic parameter of dependence between T_g and M_n.

In addition, DSC results depicted in table 9 are also consistent with the Fox-Flory relationship and demonstrate that glass transition temperature increases with increasing molecular weight. Thus, one can say that the mobility of macromolecular chains decrease when the polymer molecular weight increases. In fact, higher macromolecular chains length induces higher disentanglement rate values. Consequently, on the one hand, the required temperature value for the hemispheres surfaces softening is evidently more important. On the other hand, the force necessary to pull-out and disentangle longer macromolecular chains is significantly higher comparing to shorter macromolecules. Thus, one can say that hemispheres with shorter molecular chains entail attrition under softer conditions.

Table 9. Polystyrene bulk glass transition temperatures, T_g, determined by DSC measurements.

PS name	Tg (°C)
PS-26	94
PS-118	102
PS-245	104
PS-385	108

To conclude this section, one can say that a correlation between polymer properties and attrition appearance critical conditions could then be proposed. In fact it has been shown that greater molecular weights (with higher glass transition temperature) need higher critical speed values to generate particles damage. That means that intrinsic viscoelastic characteristics of polymers determine the attrition appearance conditions. Glass transition temperature is a critical condition to reach in order to melt the polymer surface: sufficient normal force and speed are necessary to reach this thermal value (*heat generated by friction*). Shear-thinning behavior will also inflict a critical speed able to disentangle chains and generate polymer damage.

- Effect of the wall chemistry on attrition appearance

The role interface chemistry is investigated by comparing experimental results for the attrition appearance when polystyrenes rub against hydrophobic walls (methyl-terminated) and hydrophilic walls (hydroxyl-terminated). Results reported in table 10 show that attrition appears for the same sliding velocity (25 $mm.min^{-1}$) and a minimum normal force equal to 5N and 7N, respectively, for PS-26 and PS-118. Therefore, it clearly appears that the applied normal force values required to induce PS attrition appearance onto methyl-terminated walls are much higher comparing to values required to observe the same phenomenon onto hydroxyl-terminated silicon walls. Moreover, it is important to note that no PS-245 and PS-385 attrition has been observed during experiments onto hydrophobic walls.

Table 10. Critical conditions (colored cells) for attrition appearance onto hydrophobic wall, for 4 PS ($\Delta\mu = \pm$ 0.02)

V ($mm.min^{-1}$)	0.1				1				10				25			
F_n (N)	1	3	5	7	1	3	5	7	1	3	5	7	1	3	5	7
PS name																
PS-26	0.11	0.13	0.13	0.14	0.13	0.14	0.16	0.17	0.14	0.16	0.17	0.17	0.15	0.17	0.19	0.2
PS-118	0.1	0.13	0.12	0.13	0.12	0.13	0.15	0.15	0.13	0.15	0.16	0.17	0.15	0.16	0.17	0.18
PS-245	0.09	0.12	0.13	0.13	0.12	0.12	0.13	0.14	0.13	0.14	0.13	0.16	0.14	0.14	0.16	0.18
PS-385	0.1	0.11	0.12	0.13	0.11	0.12	0.13	0.13	0.13	0.13	0.14	0.15	0.14	0.15	0.15	0.16

AFM images showed a fluidified aspect of PS that consists of liquid-like film structures onto hydrophilic surfaces, and liquid-like droplets structure onto hydrophobic surfaces (*see figure 7*).

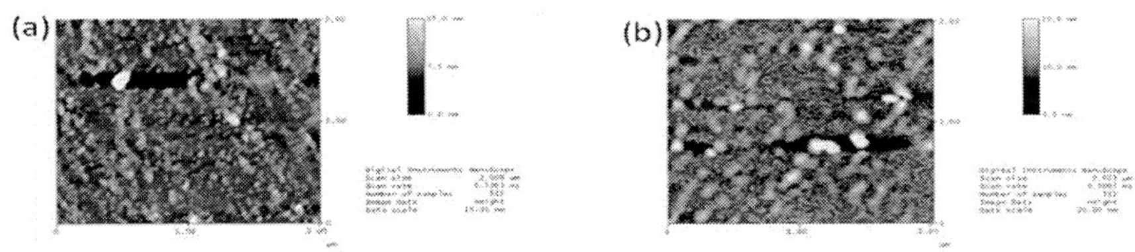

Figure 7. *AFM tapping mode images of PS-26 attrition appearance. (a)* Liquid-like film structures onto hydrophilic walls and *(b)* liquid-like droplets structure onto hydrophobic walls.

The fluidification of polystyrenes during friction could be induced by a coupled effect of the temperature and the shear-thinning. The high surface energy of hydrophilic wafers avoids dewetting. However, during friction between polystyrene and hydrophobic surface, the transfer had droplet structure, induced by a dewetting mechanism. On the basis of experimental observations, we propose a schematic, shown in Figure 8, on the spreading and the dewetting of polystyrenes onto respectively hydrophilic and hydrophobic wafers.

Indeed, on hydrophilic surfaces and for a 5 N applied normal force , the PS-26 transfer appears when the sliding speed value was about 0.1 $mm.min^{-1}$, whereas the PS-118 attrition appearance needed a higher applied normal load of about 7 N. However, on hydrophobic surfaces, the attrition appears for the same applied load comparing to values obtained on hydrophilic walls (*5 N and 7 N for PS-26 and PS-118, respectively*) but the sliding speed generating the attrition process of both polystyrenes (PS-26 and PS-118) was higher and equal to 25 $mm.min^{-1}$.

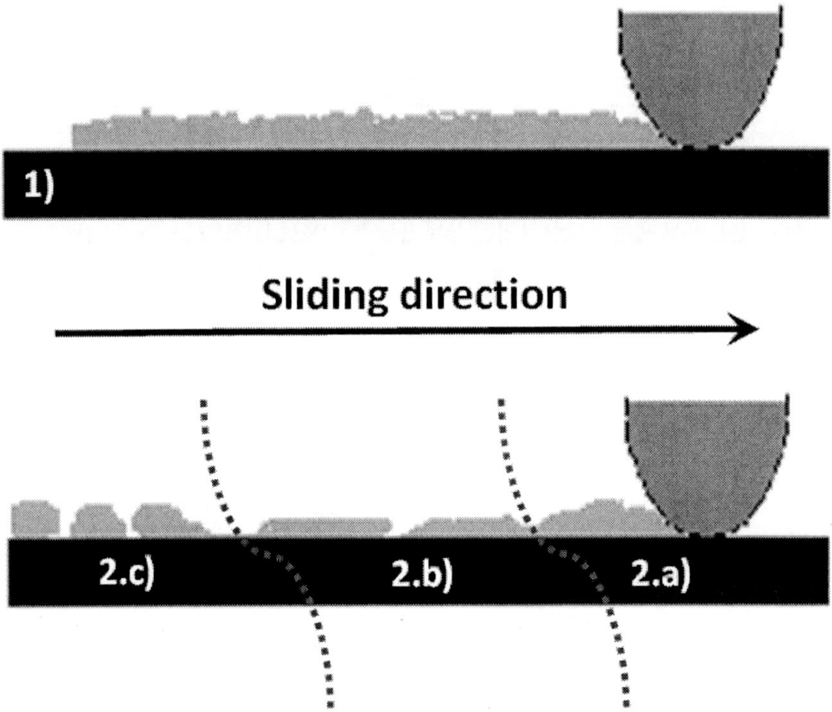

Figure 8. 1) A schematic depicting the polystyrene spreading onto hydrophilic surface. 2) A schematic depicting the dewetting of polystyrene onto hydrophobic surface. Each figure explains a) a polystyrene transfer layer is formed during friction, b) creation of dewetted polystyrene holes with exposure of the hydrophobic surface, c) mergence of dewetted polystyrene into droplets.

It appears that the normal force needed for PS-26 attrition appearance is always lower than that required for the PS-118, whatever the wall surface energy. This implies that PS-26 softening temperature is more easily reached. On the other hand, the force required to pull out shorter macromolecular chains (PS-26) and entail the attrition appearance is therefore lower.

Moreover, on the basis of reported results in table 6 and table 10, one can say that the surface chemistry plays an important role through the friction coefficient values change and through the sliding velocity critical values needed for the attrition appearance of PS-26 and PS-118 (*0.1* $mm.min^{-1}$ *and 25* $mm.min^{-1}$ *on hydrophilic and hydrophobic walls, respectively*).

3.3. Discussion of Adhesion and Rubbing at the Nanometric Scale

AFM probe tip surfaces have been chemically treated (Piranha and self-assembly methods, *see 2.2- specimen preparation*) to obtain hydrophilic and hydrophobic tips with respectively hydroxyl-terminated and methyl-terminated surfaces. Nano-adhesion between AFM tips and spin-cast polystyrene (PS-26, PS-118, PS-245 and PS-385) onto silicon surfaces has been firstly quantified owing to force distance curves. Table 11 depicts the deflection and the corresponding nano-adhesion force for 4 polystyrenes and different AFM tips.

Table 11. Deflection and nano-adhesion force between spin-cast polystyrenes surfaces and chemically modified AFM tips surfaces.

	Deflection (nm)				nano-adhesion force (nN)			
	PS-26	PS-118	PS-245	PS-385	PS-26	PS-118	PS-245	PS-385
Tip								
Si3N4-CH3	100 (±10)	115 (±10)	145 (±10)	175 (±10)	17 (±2)	19 (±2)	24 (±2)	29 (±2)
Si3N4-OH	450 (±30)	560 (±30)	630 (±30)	800 (±30)	77 (±5)	96 (±5)	108 (±5)	137 (±5)

Nano-adhesion results show higher interfacial interactions between the hydroxylated tips and polymer surfaces comparing to interactions between chlorosilane grafted tips and PS surfaces. A good correlation between surface energy and nano-adhesion is then observed. Moreover, the higher molecular weight polystyrene exhibits a greater adhesion, however, the difference between polymers is more significant in the case of hydroxylated tips. This phenomenon is may be due to the higher amount of adsorbed water between polystyrene surfaces and the hydroxylated tips, leading to the formation of a water meniscus and stronger interactions between both surfaces. These interactions could be even increased via acid–base bonding between the π electrons of the phenyl group of polystyrene and the hydroxylated tip surface (polar groups present on the tip, e.g. water molecules adsorbed onto the tip) but also be induced by the formation of water meniscus bridges.

Nano-friction experiments have been carried out for different applied normal forces and sliding speeds. In this section, the influence of the normal load and the sliding speed were analyzed. The dependence of the frictional force (TMR) on velocity is illustrated in figure 9, for an applied normal load about 0 N. The finite friction at zero normal force proves that there is nonzero adhesion between tips and polystyrene surfaces. As mentioned earlier, when the tip and the polymer film surface are brought into contact in air, the two surfaces will adhere to each other because of Van der Waals interactions and the Laplace pressure of the capillary condensed water at the periphery of the contact zone. It is important to note that under such conditions friction of the tip does not induce any scratch or damage of the film. This point has been verified by imaging the surface after friction measurements.

Sliding velocity effect. We observed that, with hydrophilic AFM tips, the friction force (TMR) increased with the sliding velocity increase (*see Figure 9*), whereas, with hydrophobic probes, the TMR remained more or less constant.

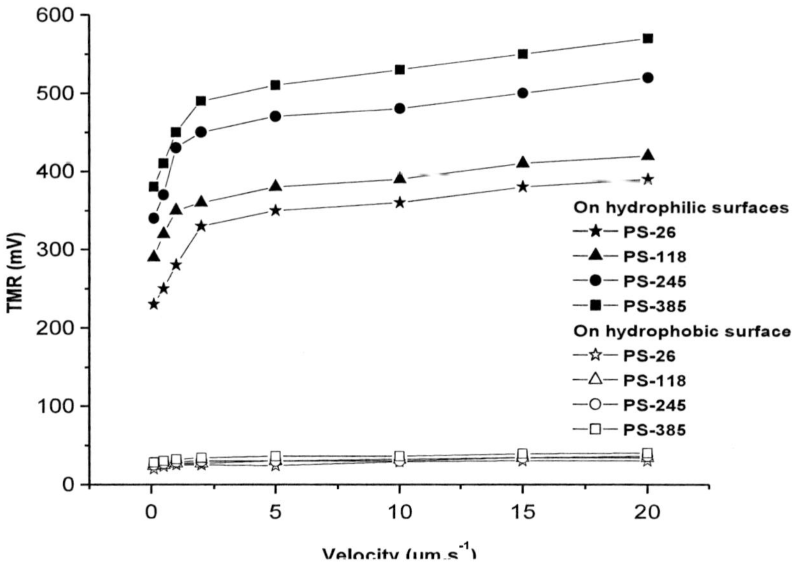

Figure 9. Friction force (TMR) versus sliding velocity for PS-26, PS-118, PS-245 and PS-385 rubbed against hydrophilic and hydrophobic walls.

The increase of friction with sliding velocity represents a rather common behavior for polymers [35–40]. A similar logarithmic increase of nano-friction with sliding velocity has been also observed for polystyrene films by Sills and Overney [3]. Various phenomenological models based on viscoelastic dissipation and relaxation times have been proposed in the literature. There, viscoelastic dissipation is supposed to occur mainly due to motion, disentanglement (and possibly scission) of chains during sliding of the tip. A higher sliding velocity leads to more dissipative losses caused by faster molecular motions, with reference to a characteristic relaxation time associated to chain mobility. Another interpretation, which is complementary to this hypothesis, is given by an Eyring-type model [43] of a thermally activated process. In this case, the tip has to overcome a series of energy barriers when sliding across the polymer molecules at the surface. Thermal fluctuation can activate this process, with a ''rate of success'' being higher for slower sliding velocities [4].

Moreover higher friction forces values recorded with hydrophilic tips could be due to the presence of a higher amount of adsorbed water (comparing to hydrophobic tips) between the polystyrene surface and the tip, leading to the formation of a water meniscus and stronger interactions between both materials, may induce more deformations of the conformations of the polymer chain at the surface and consequently to much more dissipations during the slide.

Applied normal load effect. The dependence of frictional force upon load was studied for 1 μm.s^{-1} sliding speed. Figure 10 illustrates the variation of the TMR versus the applied normal force. Figure 10 shows that the friction force increases with the applied normal load roughly in a linear manner, whatever the tip and polystyrene. This linear increase of friction force proves that the friction coefficient is independent on the applied normal force.

The linear increase of friction forces can be modelled by a modified form of the generalized Amonton's law where the friction force (F_f) is given by:

$$F_f = F_0 + \mu . F_N \qquad \text{Eq.(8)}$$

Where μ is the coefficient of friction defined as $\mu = \delta F_f/\delta F_N$, F_N the applied normal force and F_0 is the friction force when F_N is zero. The valuc of F_0 is thought to correlate but not equal to adhesion forces obtained from force-distance curve measurements.

Furthermore, it is important to note that friction forces (TMR values) are higher for the hydrophilic tips. This higher friction is probably induced by the higher adhesion between polystyrene surfaces and hydrophilic tips.

Regardless of our experimental conditions, i.e. with hydrophilic or hydrophobic tips, the friction force increases with the normal force increase for 4 PS. The present comparative results evidence that the AFM probe surface energy affect dramatically the friction force value generated during the rub. Indeed, with hydrophilic tips, the friction force increase with the sliding speed increase. However, with hydrophobic tips, we observed that the sliding velocity increase slightly affect the friction force value. Furthermore, nano-friction results are in good correlation with macro-friction results, which have shown higher friction force values of the polystyrene-hydrophilic surfaces system compared to the polystyrene-hydrophobic surfaces system.

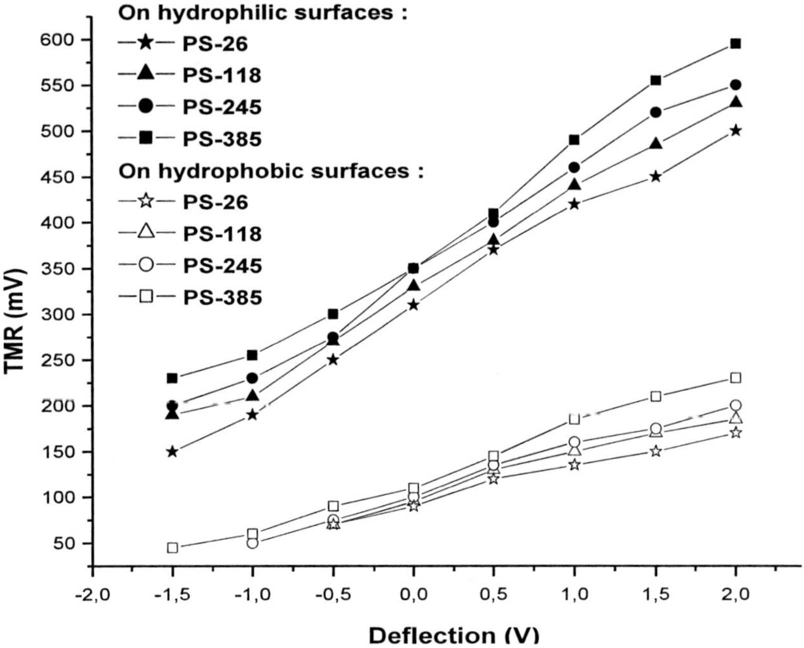

Figure 10. Friction force (TMR) versus applied normal force (Deflection set point) for PS-26, PS-118, PS-245 and PS-385 rubbed against hydrophilic and hydrophobic walls.

CONCLUSION

The dependence of the PS hemispheres attrition phenomenon on experimental and structural parameters has been analyzed in detail for a linear sliding conformation. In fact, we showed that a critical value of normal force and velocity are necessary to activate attrition mechanisms. These findings could help in the understanding of the phase density and the velocity effect on particles damage during chemical engineering processes (e.g. dense-phase conveying). Moreover, we proved that not only normal stresses and particles velocity are important in such applications but that polymers molecular weights play an important role as well. In fact, we established a correlation between polymer properties and critical conditions of attrition appearance. For instance, it was found that the glass transition temperature is a key factor in the activation of attrition mechanism of PS hemispheres. Thus one can corroborate that, materials intrinsic properties govern mechanisms leading to the attrition.

Furthermore, nanometric-results are in good correlation with macroscopic-results. In fact, results showed that friction force values registered for the (polystyrene-hydrophilic surfaces system) are higher than those registered for the (polystyrene-hydrophobic surfaces system). Moreover, nano-adhesion measurements confirmed the relation between the friction force and interfacial interactions. Consequently, it clearly appears that the nanometric approach is necessary so as to well understand the frictional behavior and therefore its repercussion on the attrition appearance.

ACKNOWLEDGMENT

This work was supported in part by the "Institut Supérieur des Sciences Appliquées et de Technologie" of Gabès (*Tunisia*) and the "Service de Physique et de Chimie des Surfaces et Interfaces du Commissariat à l'Energie Atomique" of Saclay (*France*).

REFERENCES

[1] Liu, S.-P.; Huang, I.-J.; Chang, K.-C.; Yeh, J.-M. *J. App; Poly. Sci.* 2010, 115, 288–296.

[2] Heinrich, S.; Peglow, M.; Ihlow, M.; Henneberg, M.; Mörl, L. *Chem Eng.Sci.* 2002, 57 (20), 4369–4390.

[3] Kalman, H.; Rabinovich, E.; *Chem. Eng. Sci.* 2008, 63, 3466–3473.

[4] Bridgwater, J. *Chem. Eng. Sci.* 1995, 50 (24), 4081–4089.

[5] Frye, L.; Peukert, W. *Powder Technol.* 2004, 143–144, 308–320.

[6] Ghadiri, M.; Zhang, Z. *Chem. Eng. Sci.* 2002, 57 (17),3659–3669.

[7] Pitchumani, R.; Meesters G. M. H.; Scarlett B. *Powder Technol.* 2003, 130(1–3), 421–427.

[8] Hu, G.-H.; Hoppe, S.; Feng, L.-F.; Fonteix, C. *Chem. Eng. Sci.* 2007, 62, 3528–3537.

[9] Cheong, Y. S.; ADAMS, M. J.; Hounslow, M. J.; Salman A. D. Trans IChemE, Part A, *Chemical Engineering Research and Design* 2005, 83(A11), 1276–1282.
[10] Cheong, Y.S.; Adams, M.J.; Hounslow, M.J.; Salman, A.D. *Powder Technol.* 2009, 189, 365–375.
[11] Kalman, H.; Goder, D. *Advanced Powder Technol.* 1998, 9 (2), 153–167.
[12] J. F. Richardson, J. H. Harker and J. R. Backhurst, Chemical Engineering, Volume 2, Fifth edition, *Particle Technology and Separation Processes*, Butterworth-Heinemann, imprint of Elsevier Science, 2002.
[13] Prasher, C. L. Crushing and Grinding Process Handbook (Wiley, 1987).
[14] Bemrose, C.R.; Bridgewater, J. A. *Powder Technol.* 1987, 49, 97–126.
[15] Ahmadian, H.; Ghadiri, M. *Advanced Powder Technol.* 2007, 18(1), 53–67.
[16] Reynolds, G.K.; Fu, J.S.; Cheong, Y.S.; Hounslow, M.J.; Salman, A.D. *Chem. Eng. Sci.* 2005, 60, 3969–3992.
[17] Scala, F.; Montagnaro, F.; Salatino, P. *Energy & Fuels* 2007, 21, 2566-2572.
[18] Couroyer, C.; Ning, Z.; Ghadiri, M. *Powder Technol.* 2000, 109, 241–254.
[19] Samimi, A.; Moreno, R.; Ghadiri, M. *Powder Technol.* 2004, 143–144, 97– 109.
[20] Ghadiri, M.; Ning, Z.; Kenter, S. J.; Puik, E. *Chem. Eng. Sci.* 2000, 55, 5445-5456.
[21] Ghorbal, A.; Ben Arfi, R.; Bistac, S.; Brogly, M. *Chem. Phys.Lett.* 2007, 443, 352–355.
[22] Ren, S.-L.; Yang, S.-R.; Wang, J.-Q.; Liu, W.-M.; Zhao, Y.-P. *Chem. Mater.* 2004, 16, 428.
[23] Dao, T. T.; Archer, L. A. *Langmuir* 2001, 17, 4042.
[24] Maeda, N.; Chen, M. N.; Tirrell, M.; Israelachvili, J. N. *Science* 2002, 297, 379.
[25] Charlier, J.; Ghorbal, A.; Grisotto, F.; Clolus, E.; Palacin, S.; Main strategies to direct localized organic grafting on conducting and semiconducting substrates; In *New Nanotechniques*; Malik, A., Rawat, R. J., Eds.; Nova Science Publishers: 2009; Vol. chap 9, p 319-350 (ISBN: 978-1-60692-516-4).
[26] Palacin, S.; Bureau, C.; Charlier, J.; Deniau, G.; Mouanda, B.; Viel, P. *Chem. Phys. Chem.* 2004, 5, 1468.

[27] Ghorbal, A.; Grisotto, F.; Laudé, M.; Charlier, J.; Palacin, S. *J. Colloid Interface Sci.* 2008, 328 (2), 308-313.
[28] Ulman, *A. Chem. Rev.* 1995, 96, 1533.
[29] Israelachvili N. J., *Molecular and surface forces*, 2nd edition, Academic Press New-York, 1991.
[30] Sellers, H.; Ulman, A.; Shnidman, Y.; Eilers, J. E. *J. Am. Chem. Soc.* 1993,115, 9389-9401.
[31] Tripp C. P.; Hair M. L. *Langmuir* 1992, 8, 1120-1126.
[32] Allara, D. L.; Parikh, A. N.; Rondelez, F. *Langmuir* 1995,11, 2357-2360.
[33] Lee, W.; Lee, H.; Cha, J.; Chang, T.; Hanley K. J.; Lodge, T. P. *Macromolecules* 2000, 33, 5111-5115
[34] Ndoni, S.; Papadakis, C. M.; Bates, F. S.; Almdal, K. *Rev. Sci. Instrum.* 1995, 66, 1090.
[35] Haidara, H.; Vonna, L.; Schultz, J. *Langmuir* 1996, 12, 3351- 3355.
[36] Lo, Y.-S.; Huefner, N. D.; Chan, W. S.; Dryden, P.; Hagenhoff, B.; Beebe, T. P., Jr. *Langmuir* 1999; 15(19); 6522.
[37] Haidara, H.; Mougin, K. Liquid Ring Formation from Contacting, Nonmiscible Sessile Drops, *Langmuir* 2005, 21, 1895-1899
[38] Mougin, K.; Haidara, H., Castelein. G.; *Colloids Surf. Physicochem. Eng. Aspects*. 2001;193:231–7.
[39] Walsh, C. B.; Franses, E. I. *Thin Solid Films* 2003, 429, 71–76.
[40] Fowkes, F. M. *J. Phys. Chem.* 1962, 66, 382.
[41] Fowkes, F. M. *J. Phys. Chem.* 1963, 67, 2538–2541.
[42] Fowkes, F. M. *Industrial and Engineering Chemistry* 1964, 56 (12).
[43] Fowkes, F. M. *J. Phys. Chem.* 1968, 72, 3700.
[44] Owens, D. K; Wendt, R. C. *J. Appl. Polym. Sci.* 1969, 13, 1741.
[45] van Oss, C.J.; Good, R.J.; Chaudhury, M. K. *J. Colloid.Interface Sci.* 1986, 111, 378.
[46] van Oss, C.J.; Giese, R.F.; Wentzek, R.; Norris, J.; Chuvilin, E.M. *J. Adh. Sci. Tech.* 1992, 6, 503.
[47] van Oss, C.J.; Giese, R.F.; Wu, W. *J. Adhesion* 1997, 63, 71.
[48] Binning, G.; Quate, C. F.; Gerber C. *Phys. Rev. Lett.* 1986, 56, 930–933.
[49] Zhong, Q.; Innis, D.; Kjoller, K.; Elings. V. *Surf. Sci. Lett.* 1993, 290, L688–92.
[50] Bhushan, B. Springer Handbook of Nanotechnology, Spinger-Verlag Berlin Heidelberg New York, 2004
[51] Noel, O.; Brogly, M.; Castelein, G.; Schultz, J. *European Polymer Journal* 2004, 40, 965–974.

[52] Carpick, R. W.; Salmeron, M. *Chem. Rev.* 1997, 97, 1163–1194.
[53] Bhushan B.: Handbook of Micro/Nanotribology, 2nd edn. (CRC, Boca Raton 1999).
[54] Bhushan, B.; Dandavate, C. *J. Appl. Phys.* 2000, 87, 1201–1210
[55] Clear, S. C.; Nealey, P. F. *J. Colloid Interface Sci.* 1999, 213, 238–250.
[56] Lee, D. H.; Oh, T.; Cho, K. *J. Phys. Chem. B* 2005, 109, 11301-11306
[57] Cleveland, J. P.; Manne, S.; Bocek, D.; Hansma, P. K. *Rev. Sci. Instrum.* 1993, 64, 403-405.
[58] Hutter, J. L.; Bechhoefer, J. *Rev. Sci. Instrum.* 1993, 64, 1868-1873.
[59] Senden, T. J.; Ducker, W. A. *Langmuir* 1994, 10, 1003-1004.
[60] Levy, R.; Maaloum, M. *Nanotechnology Bristol* 2002, 13, 33-37.
[61] Torii, A.; Sasaki, M.; Hane, K.; Okuma, S. *Meas. Sci. Technol.* 1996, 7, 179-184.
[62] Green, C. P.; Lioe, H.; Cleveland, J. P.; Proksch, R.; Mulvaney, P.;. Sader, J. E. *Rev. Scientific Instr.* 2004, 75, 1988-1996.
[63] Sader, J. E.; Chon, J. W. M.; Mulvaney, P. *Rev. Scientific Instr.* 1999, 70, 3967-3969.
[64] Ghatak, A.; Vorvolakos, K.; She, H.; Malotky, D. L.; Chaudhury, M. K. *J Phys Chem B* 2000, 104, 4018.
[65] Persson B. N. J. Surf Sci 1998, 401, 445. 17. Chen, N.; Maeda, N.; Tirrell, M.; Israelachvili, J. *Macromolecules* 2005, 38, 3491.
[66] Ren, S.-L.; Yang, S.-R.; Wang, J.-Q.; Liu, W.-M.; Zhao, Y.-P. *Chem. Mater.* 2004, 16, 428.
[67] Dao, T. T.; Archer, L. A. *Langmuir* 2001, 17, 4042.
[68] Liu, Y.; Evans, D. F.; Song, Q.; Grainger, D. W. *Langmuir* 1996, 12, 1235.
[69] Thomas, E. L. *Langmuir* 1996, 12, 4557.
[70] Brule , B.; Kausch, H. H.; Monnerie, L.; Plummer, C. J. G.; Halary, J. L. *Polymer* 2003, 44, 1181.
[71] Maeda, N.; Nianhuan Chen, M.; Tirrell, M.; Israelachvili, J. N. *Science* 2002, 297, 379.
[72] Luengo, G.; Schmitt, F.-J.; Hill, R.; Israelachvili, J. *Macromolecules* 1997, 30, 2482.
[73] Wallace, W. E.; Fischer, D. A.; Efimenko, K.; Wu, W.-L.; Genzer, J. *Macromolecules* 2001, 34, 5081.
[74] Hu, H.; Carson, G. A.; Granick, S. *Phys. Rev. Lett.* 1991, 66, 2758.
[75] Luengo, G.; Israelachvili, J.; Granick, S. *Wear* 1996, 200, 328.

[76] Israelachvili, J.; Berman, A. D. In CRC Handbook of Micro/Nanotribology, 2nd ed.; Bhushan, B., Ed.; CRC Press: Boca Raton, FL, 1999; Ch. 9, pp 372–432.
[77] Xie, F.; Zhang, H. F.; Lee, F. K.; Du, B.; Tsui, O. K. C.; Yokoe, Y.; Tanaka, K.; Takahara, A.;
[78] Kajiyama, T.; He, T. *Macromolecules* 2002, 35, 1491.
[79] Gotsmann, B.; Durig, U. *Langmuir* 2004, 20, 1495.
[80] Tillman, N.; Ulman, A.; Schildkraut, J. S.; Penner, T. L. *J. Am. Chem. Soc.* 1988,110,6136.
[81] Ulman, A. *J. Mater. Educ.* 1989, 11, 205.
[82] Millan, M.D.; Locklin, J.; Fulghum, T.; Baba, A.; Advincula, R.C. *Polymer* 2005, 46, 5556.
[83] G. Socrates, Infra Red Characteristic Group Frequencies, Wiley & Sons, NY, 1980.
[84] Graessley, W. W. *Adv. Polym. Sci.* 1974, *16*, 1.
[85] Kagata, G.; Gong, J. P.; Osada, Y. *J. Phys. Chem. B.* 2002, 106(18), 4596.
[86] Aoike, T.; Uehara, H.; Yamanobe, T.; Komoto, T. *Langmuir* 2001, *17*(7), 2153.
[87] Suda, H. *Langmuir* 2001, *17*(20), 6045.
[88] *Tabor, D; British Journal of Applied Physics 1950-1967* 1955, 6(3), 79-81.
[89] *Minato, K.; Takemura, T. Japanese Journal of applied physics* 1967, 6(6).
[90] Tomlinson Fort, Jr. *J. Phys. Chem.* 1962, 66(6), 1136.
[91] Ren, S.-L.; Yang, S.-R.; Wang, J.-Q.; Liu, W.-M.; Zhao, Y.-P. *Chem. Mater.* 2004, 16(3), 428.
[92] Bahadur, S. *Wear* 2000, 245(1-2), 92.
[93] Berthoud, P.; G'Sell C.; Hiver, J.-M. *J. Phys. D: Appl. Phys.* 1999, 32, 2923–2932.
[94] Ghorbal, A.; Schmitt, M.; Bistac, S. *Journal of Polymer Science: Part B: Polymer Physics* 2006, 44, 2449–2454.
[95] Léger, L; Hervet,H.; Bureau,L. *Comptes Rendus Chimie* 2006, 9(1) 80-89.
[96] *Kerle, T.; Lin, Z.; Kim, H.-C.; Russell, T. P.; Macromolecules* 2001; 34*(10); 3484-3492.*

In: Polystyrene: Properties, Performance... ISBN: 978-1-61209-117-4
Editor: James E. Gray

Chapter 2

BIODEGRADABILITY OF POLYSTYRENE THAT CONTAINS *N*-BENZYL-4-VINYLPYRIDINIUM CHLORIDE IN THE MAIN CHAIN

*Nariyoshi Kawabata**
Kyoto Institute of Technology, Kyoto, Japan
University of Shiga Prefecture, Shiga, Japan

ABSTRACT

In protecting the environment from pollution by waste polymers, it is indispensable to make synthetic polymers biodegradable by modification of the chemical structure without severe damage of utility. Biodegradation of polystyrene is extremely difficult, but copolymer of styrene with *N*-benzyl-4-vinylpyridinium chloride, PSt-*co*-BVP(Cl), is degradable when treated with activated sludge in soil. Biodegradation of the copolymer that contained 49.1-mol % BVP(Cl) followed first order kinetics and half-life was 5.6 days when 1.0 or 0.5 g/kg of polymer sample was used. Biodegradation of the copolymer with reduced BVP(Cl) content did not follow first order kinetics under the conditions,

* Resident address: 4-5-6 Kitahorie, Nishi-ku, Osaka 550-0014, Japan. Kyoto Institute of Technology, Matsugasaki, Kyoto 606-8585, Japan, University of Shiga Prefecture, 2500 Hassaka-cho, Hikone, Shiga 522-8533, Japan. Tel.: +81-6-6539-0390; fax.: +81-6-6539-0490. E-mail address: bfalq308@cwo.zaq.ne.jp.

but appeared to follow the kinetics when a sufficiently small amount of polymer sample was used. Under ultimate conditions, half-life of the copolymer that contained 10.6-mol % BVP(Cl) was estimated 12.5 days. Half-life of the copolymer that contained 5-mol % BVP(Cl) was assumed 30-40 days. Insertion of 5-mol % BVP(Cl) appeared sufficient for making the copolymer substantially biodegradable if we do not expect exceptionally rapid degradation. Random scission of the main chain predominated over uniform scission from the end of polymer chain in the biodegradation of PSt-*co*-BVP(Cl). Cleavage of the main chain at BVP(Cl) appeared predominant over that at oligo-styrene portion. There are a variety of noticeable utilities of the modified polystyrene due to the strong affinity of BVP(Cl) with microbial cells and viruses.

INTRODUCTION

Biodegradable polymers have received increasede attention [1-7], but developments of these polymers are mainly focused on the substitution of non-biodegradable polymers for biodegradable polymers, regarding their utility as secondary importance. There are many useful polymers with poor biodegradability. For example, polystyrene has a variety of utilities but its biodegradability is extremely poor. Therefore, it is indispensable to make these non-biodegradable polymers biodegradable by partial modification of the chemical structure without severe damage of their utility.

Biodegradable polymers are not limited to natural polymers such as derivatives of cellulose, starch-based polymers, chitin and polyhydroxy-alkanoate produced by bacteria. Synthetic polymers such as homopolymers and copolymers of lactic acid, polyethylene adipate, polytetramethylene adipate, polycaprolactone and other aliphatic polyesters are biodegradable. The main chain of these synthetic polymers is consisted of ester group, and is decomposable by hydrolysis reaction catalyzed by hydrolase in the natural environment.

However, vinyl polymers having carbon-carbon bond as the main chain are poorly biodegradable except for poly(vinyl alcohol) [8]. In this case, oxidative degradation of carbon-carbon bond of the main chain is indispensable. In this biodegradation, hydrophilicity of poly(vinyl alcohol) is favorable whereas hydrophobicity of polystyrene is unfavorable.

Nevertheless, low molecular weight polymers and oligomers of vinyl compounds are biodegradable dissimilar to the corresponding high molecular weight polymers [9-13]. This fact suggests that high molecular weight vinyl polymers become biodegradable by the insertion of highly biodegradable chemical units into the main chain. That is, synthesis of biodegradable vinyl polymers can be realized by the connection of biodegradable oligomers using highly biodegradable chemical units. For example, Matsumura et al. [14] made poly(sodium acrylate) biodegradable by insertion of vinyl alcohol into the main chain.

High molecular weight polyethylene is poorly biodegradable, but the corresponding low molecular weight compounds such as dotriacontane, $CH_3(CH_2)_{30}CH_3$, and other long-chain alkanes are biodegradable [9]. High molecular weight poly-1,4-butadiene is poorly biodegradable, but oligomers of butadiene are biodegradable, and about 40% of polybutadiene having a number average molecular weight 2,350 (degree of polymerization 44) was degraded by a microbial strain capable of utilizing 1,4-type polybutadiene as a sole source of carbon in 5 days [10]. High molecular weight poly-1,4-isoprene is poorly biodegradable, but oligomers of isoprene are biodegradable, and about 60% of *cis*-1,4-polyisoprene having a number average molecular weight 940 (degree of polymerization 14) was degraded by a microbial strain in 4 days [11]. The microbial strain was isolated from soil by enrichment cultures and repeated isolations on plate cultures. High molecular weight polyacrylonitrile is poorly biodegradable, but trimer of acrylonitrile was utilized by bacteria [12].

In contrast, however, biodegradation of even low molecular weight oligomers of styrene was reported extremely difficult [13]. Unsaturated dimer of styrene, 1,3-diphenyl-1-butene, was utilized by *Alcaligenes* sp. isolated from soil, and rapidly disappeared from the culture medium. This bacterial strain also utilized cyclic dimer of styrene, 1-methyl-3-phenylindane. However, no degradation of trimers of styrene by the bacterial strain was detected.

On the other hand, we accidentally discovered proliferation of *Escherichia coli* on the surface of cross-linked poly(*N*-benzyl-4-vinylpyridinium bromide), cross-linked PBVP(Br), during a study on a new method useful for water disinfection [15]. This phenomenon clearly indicated digestion of cross-linked PBVP(Br) by *E. coli*, because there was no nutritive material for microbes other than the polymer. However, we did not become aware of the exceeding importance of this phenomenon at that time.

After about 10 years, we were shocked by the violent digestion of cross-linked PBVP(Br) by activated sludge when placed in a system for continuous aerobic treatment of artificial sewage [16]. We perceived that cross-linked PBVP(Br) was an excellent nutrient for microbes. Based on this recognition, we tried to make vinyl polymers biodegradable by the insertion of *N*-benzyl-4-vinylpyridinium chloride, BVP(Cl), into the main chain as a more strongly biodegradable chemical unit than vinyl alcohol that was used by Matsumura et al. [14].

We made polystyrene biodegradable by this methodology [17]. We were encountered with an unexpected phenomenon. In biodegradation of copolymer of styrene with BVP(Cl), PSt-*co*-BVP(Cl), biodegradation of oligo-styrene portion was not limited to dimer, and larger oligo-styrene portion was degraded. BVP(Cl) is not only a biodegradable unit useful to connect oligo-styrene, but also stimulates microbes to degrade oligo-styrene connected to BVP(Cl).

This chapter surveys biodegradation of PSt-*co*-BVP(Cl) in molar ratios 1:1 to 8:1 by the treatment with activated sludge in soil [17].

Indispensable Quality Required for Biodegradable Polymer

Quality Required for Excellent Biodegradability

According to the majority of public recognition, biodegradable polymers are biologically decomposable polymers after use, and decomposability is sufficient for the protection of the environment from pollution by waste polymers. However, this recognition is quite one-sided and disorders the development of synthetic polymers having excellent biodegradability.

Partial decomposition is not sufficient. It is indispensable that waste polymers are perfectly participated in circulation system of materials in the natural ecological environment.

From this point of view, violent digestion of cross-linked PBVP(Br) by activated sludge is an epoch-making discovery [16]. This fact proves nutritive worth of the polymer for microbes contained in activated sludge. Enhancement of nutritive worth is an indispensable subject for biodegradable polymers. Nitrogen included in BVP(Cl) is an indispensable nutritive element for microbes. Poly(vinyl alcohol) is biodegradable [8] but does not contain nitrogen. Nutritive worth of BVP(Cl) is obviously much higher than that of vinyl alcohol.

The introduction of carboxyl, hydroxyl, amino, amido, pyridyl and other hydrophilic groups lets hydrophobic polymers wear hydrophilicity, and assist biodegradation of the polymers. From this point of view, BVP(Cl) is strongly hydrophilic and very effective for this purpose.

Quality Required for Test Microbes in Charges of Biodegradation

In the case where a polymer showed biodegradability by using a specific bacterial strain, the result does not always certificate biodegradation of the subject polymer in the natural environment. In the case where a polymer showed biodegradability by using activated sludge obtained from an aeration tank of sewage works, the biodegradability is more reliable, because activated sludge includes a variety of microbes and forms an ecological system of microbes. Therefore, we principally used activated sludge as microbes in charges of biodegradation.

Durability and Biodegradability of Synthetic Polymer

Biodegradability is important after use, whereas durability is important during use. Therefore, development of a highly durable synthetic polymer shows a tendency to confront with that of a highly biodegradable synthetic polymer. The development of a highly durable polymer is essentially incompatible with that of a highly biodegradable polymer.

In the case where durability is definitively important, we have to recover used polymers completely, and decompose or repeatedly use them adequately, in order to prevent pollution of the natural ecological environment by waste polymers. It is not desirable to treat used polymers as useless materials. We should treat used polymers as valuable resources, and use them repeatedly. There is no useless waste in the natural ecological environment.

On the other hand, in the case where we expect complete decomposition of the used polymers in natural ecological environment, excellent biodegradability is indispensable for these synthetic polymers. In this case, waste polymers should be excellent nutrients for microbes in charges of biodegradation of the polymer in the natural ecological environment.

Principal Contribution of *N*-Benzyl-4-Vinylpyridinium Chloride to Biodegradation of The Modified Polystyrene

BVP(Cl) exhibited five important contributions to the biodegradation of PSt-*co*-BVP(Cl). In the first place, extraordinarily strong biodegradability of BVP(Cl) begins biodegradation at BVP(Cl) portion of PSt-*co*-BVP(Cl). Secondly, BVP(Cl) powerfully stimulates microbes to degrade oligo-styrene connected to BVP(Cl). Thirdly, highly nutritive worth of BVP(Cl) for microbes accelerates biodegradation, and facilitates well-matched circulation of PSt-*co*-BVP(Cl) in the natural environment. Fourthly, strong affinity of BVP(Cl) with microbial cells increases the opportunity of biodegradation of PSt-*co*-BVP(Cl) in the natural environment. Finally, strong hydrophilicity of BVP(Cl) makes biodegradation of PSt-*co*-BVP(Cl) easier by relieving strong hydrophobicity of polystyrene.

Prior to the descriptions on biodegradation of PSt-*co*-BVP(Cl) by activated sludge in soil, I briefly summarize these important contributions of BVP(Cl) to the biodegradation.

Extraordinarily Strong Biodegradability

In the case of water-soluble chemical materials, the ratio of biochemical oxygen demands for 5 days, BOD_5, to total organic carbon, TOC, is an index of biodegradability, because TOC reflects substantial concentration of organic materials, whereas BOD_5 is a criterion of the concentration of biologically degradable materials in a sample solution.

BOD_5/TOC of a commercial poly(vinyl alcohol) (100% hydrolyzed, average molecular weight 14,000) was 0.033, but that of polyacrylamide (average molecular weight 309,000) was zero [18]. In a striking contrast, BOD_5/TOC of a copolymer of acrylamide with 11.0-mol % BVP(Cl) was 0.607 [18]. BVP(Cl) was proved far more biodegradable than vinyl alcohol.

Biodegradation of PSt-*co*-BVP(Cl) may begin at the portion of BVP(Cl) due to the extraordinarily strong biodegradability.

Powerful Stimulation of Microbes to Degrade the Connected Portion

Dimers of styrene are biodegradable but no degradation of trimer of styrene was detected [13]. This result prompted us to give up biodegradation of octamer of styrene. Unimaginably, however, PSt-*co*-BVP(Cl) that contained 10.6-mol % BVP(Cl) was degraded by the treatment with activated sludge in soil, and the estimated half-life under ultimate conditions was 12.5 days [17]. In this case, the average degree of polymerization of oligo-styrene portion was 8.3.

This unbelievable result proves that the function of BVP(Cl) is not only a strongly biodegradable connector of biodegradable oligo-styrene, but also a powerful stimulator of microbes to degrade the oligo-styrene portion of PSt-*co*-BVP(Cl) connected to BVP(Cl).

Highly Nutritive Worth for Microbes in Charges of Biodegradation

Proliferation of Bacteria on the Surface of Cross-Linked PBVP(Br)

We added cross-linked PBVP(Br) to suspensions of bacteria in sterilized physiological saline that did not contain nutrient materials for bacteria, but found considerable proliferation of bacteria on the polymer surface [15]. This phenomenon clearly indicated that the bacteria digested cross-linked PBVP(Br) as their nutrient, and highly nutritive worth of the polymer.

Violent Digestion of Cross-Linked PBVP(Br) by Activated Sludge

Since cross-linked PBVP(Br) captured microbial cells [15], we tried to use the polymer as a filter medium for biofilm process [16]. Fixation of biomass on the surface of a filter medium is important in a biofilm process that treats wastewater with reduced reactor volume [19-21].

We performed continuous aerobic treatments of artificial sewage by activated sludge using an apparatus filled with cross-linked PBVP(Br) prepared in the form of Raschig ring [16]. We expected effective removal of chemical oxygen demand, COD, of artificial sewage during the treatment. Contrary to this expectation, however, COD of effluent solution was 2-3 times larger than that of influent artificial sewage during the early 2-3 weeks. We attributed this considerable increase in COD to the proliferation of microbes contained in activated sludge by the digestion of cross-linked PBVP(Br) as their nutrient, because no other organic materials existed in the treatment system except for the organic components of influent artificial sewage.

Weight of cross-linked PBVP(Br) reduced vigorously during the biological treatment. For example, 80% of cross-linked PBVP(Br) disappeared during 60 days of the treatment in the case where influent COD concentration of the artificial sewage was 500 mg/L.

Violent digestion of cross-linked PBVP(Br) proved highly nutritive worth of the polymer. BVP(Br) contains nitrogen but vinyl alcohol does not contain nitrogen. Nutritive worth of BVP(Cl) is much higher than that of vinyl alcohol for microbes in charges of biodegradation.

In harmonious circulation of materials contained in synthetic polymers in the natural ecological environment, nutritive worth of the polymers for microbes in charges of biodegradation is definitively important.

Strong Bactericidal Activity of Not-Cross-Linked PBVP(Br)

It should be noted that non-cross-linked PBVP(Br) exhibits a strongly bactericidal activity, and was a match for conventional strong disinfectants such as benzalkonium chloride and chlorohexidine, making a sharp contrast to cross-linked PBVP (Br) [22]. Non-cross-linked PBVP (Br) exhibited bactericidal activity against gram-positive bacteria, but less active against gram-negative bacteria, and the activity was considerably greater than the corresponding monomer compound. Polymer effect was significant in the bactericidal activity.

In addition, it should be noted that non-cross-linked copolymer of acrylamide with 11.0-mol % BVP(Cl) did not exhibit bactericidal activity [18]. This result indicated that the bactericidal activity required several numbers of BVP(Cl) in a polymer molecule.

Strong Affinity with Microbial Cells That Increases Opportunity of Biodegradation

Strong affinity of BVP(Cl) with microbial cells increases opportunity of biodegradation. A significant aspect of this affinity is of importance to physicochemical interactions between BVP(Cl) and microbial cells than physiological action of BVP(Cl), because we observed no difference between the interaction with living bacterial cells and that with killed bacterial cells.

Capture of Bacterial Cells by Adhesion on the Surface of Cross-Linked PBVP(Br)

Cross-linked PBVP(Br) exhibited a noticeable ability to remove bacterial cells from water by capture on the polymer surface in the living state [15]. We added 10 g (wet weight) of the polymer to 20 mL of bacterial suspensions in sterilized physiological saline, and stirred the mixture at 37°C. This treatment removed 99% of viable bacteria during 2-6 h.

We passed bacterial suspensions (10^5 to 10^8 cells/mL) through a glass column (20 mm by 100 cm) packed with cross-linked PBVP(Br) at 37°C with a flow rate of 0.8-1.4 bed volumes per h. This treatment removed 97-100% of viable bacteria. This treatment removed TOC of bacterial suspensions, and proved that the polymer irreversibly captured bacterial cells. Estimated capacity of capture was 10^{10} cells per g (dry weight) of the polymer.

Captured bacteria proliferated on bacterial medium, and proved that cross-linked PBVP(Br) captured bacterial cells in the living state. Even when we killed the bacteria by heat treatment before the column studies, the amount of TOC removed by the polymer did not change. This result indicated that the capture of bacterial cells by cross-linked PBVP(Br) was not limited to living bacterial cells. Physicochemical interaction between microbial cells and cross-linked PBVP(Br) was much more important than physiological action of the polymer.

Influence of Chemical Structure on the Ability to Capture Bacterial Cells

We confirmed the strong ability of cross-linked PBVP(Br) to capture bacterial cells by equilibrium capture experiments [23]. We evaluated the capturing ability based on the removal coefficient obtained from the initial rate of the decrease of viable cell counts caused by the addition of a test polymer [15]. Cross-linked PBVP(Br) and PBVP(Cl) exhibited conspicuously higher removal coefficient than those of other synthetic polymers examined. The removal coefficient of cross-linked PBVP(Br) uniformly increased with the content of BVP(Br) in the polymer.

The removal coefficient considerably increased when the content of divinylbenzene became smaller, and indicated that a high degree of swelling in water of the polymer matrix enhanced the capturing ability. The removal coefficient also increased with the amount of 4-vinylpyridine contained in the polymer matrix, presumably due to the hydrophilicity.

Based on these observations, we concluded that the presence of BVP(Br) and BVP(Cl) was essentially important in the affinity of cross-linked

PBVP(Br) and PBVP(Cl) with bacterial cells, and hydrophilicity of the polymer matrix facilitated the ability to capture bacterial cells.

Influence of Electrostatic and Hydrophobic Interactions on the Capture of Bacterial Cells

Capturing interaction between cross-linked PBVP(Br) and bacterial cells was investigated in more detail [24]. Hydrophobicity of bacterial cells was measured by the procedure proposed by Rosenberg [25] using hexadecane, *p*-xylene, chloroform and ethyl acetate as organic solvents. We evaluated the hydrophobicity of bacterial cells based on the decrease of the absorbance of aqueous phase due to bacterial cells caused by the addition of hydrophobic organic solvents. Bacterial species used in this work were roughly classified into four groups. Class A, bacteria of high hydrophobicity, included *S. aureus*. Class B, bacteria of medium hydrophobicity, included *Bacillus* sp., *Proteus* sp., and *S. typhimurium*. Class C, bacteria of medium hydrophilicity, included *P. aeruginosa* and *Citrobacter* sp. Class D, bacteria of high hydrophilicity, included *E. coli, Enterobacter* sp., and *Klebsiella* sp.

The surface charge of bacterial cells was measured using colloid titration method [26] performed at pH 9. We expressed the surface charge as gram equivalents per bacterial cells.

Hydrophobic bacteria and hydrophilic bacteria showed distinct differences in the capturing interaction with cross-linked PBVP(Br). With hydrophobic bacteria, electrostatic interaction between the positive charge of cross-linked PBVP(Br) and negative charge of the surface of bacterial cells, as well as hydrophobic interaction between the polymer and bacterial cells, appeared important in the capturing interaction. With hydrophilic bacteria, however, experimental results suggested the importance of other factors such as solvent (water) mediated forces and hydrodynamic forces.

Strong Hydrophilicity That Assists Biodegradation

An important factor of extremely poor biodegradability of polystyrene appears to be the considerably strong hydrophobicity. Water is important in biodegradation. Non-cross-linked PBVP(Br) and PBVP(Cl) are hygroscopic, and cross-linked PBVP(Br) and PBVP(Cl) considerably swell in water when the content of divinylbenzene was low. These results indicate that BVP(Cl) and BVP(Br) have considerably strong affinity with water. Insertion of

BVP(Cl) or BVP(Br) to polystyrene improves biodegradability by lightening the strong hydrophobicity.

EXPERIMENTAL METHODS

Materials

Styrene was washed with 5 % aqueous sodium hydroxide solution followed by drying using calcium hydride. 4-Vinylpyridine was purified by distillation under a reduced pressure. Other chemicals were used without further purification. Activated sludge was obtained from Sewage Works of Hikone City, and washed with sterilized physiological saline. Test soil was obtained from the shore of Lake Biwa and washed with water using Soxhlet extraction apparatus for 48 h, and then washed with ethyl acetate using Soxhlet extraction apparatus for 15 h. After drying, the soil was sterilized by autoclaving at 121°C for 180 min. After 2 days, we repeated the autoclaving once again, and the soil was dried to constant weight. Artificial sewage was prepared according to a literature recipe [19] and used to assist the biodegradation. Peptone (6.0 g), meat extract (4.0 g), urea (1.0 g), sodium chloride (0.30 g), potassium chloride (0.14 g), calcium chloride (0.14 g), magnesium sulfate (0.10 g), and sodium hydrogenphosphate (1.0 g) were dissolved in 1.0 liter deionized water, and pH was adjusted to 8.5. COD of this undiluted solution was about 10,000 mg/L. We diluted this solution with an appropriate amount of deionized water to prepare artificial sewage of prescribed COD concentration.

Preparation of Copolymers of Styrene with *N*-Benzyl-4-Vinylpyridinium Chloride

We performed copolymerization of styrene with 4-vinylpyridine in toluene at 80°C for 24 h using 2,2'-azobisisobutyronitrile as the initiator. After polymerization, we removed volatile materials by evaporation using a rotary evaporator, and dissolved the residue that contained copolymer of styrene with 4-vinylpyridine, PSt-*co*-VP, to a mixed solvent of tetrahydrofuran and ethanol in weight ratio 1:1. We added benzyl chloride to the mixture. The amount of benzyl chloride was equimolar to 4-vinylpyridine contained in PSt-*co*-VP. We allowed the mixture to react at 70°C for 24 h. After the reaction, we removed volatile materials by evaporation using a rotary evaporator, and added hexane to the residue and isolated PSt-*co*-BVP(Cl). We dried PSt-*co*-BVP(Cl) to a constant weight under a reduced pressure at room temperature.

We ascertained the chemical structure of PSt-*co*-BVP(Cl) by elemental analyses performed at the Elementary Analyses Center of Kyoto University. Intrinsic viscosity was determined at 25°C in dimethylformamide, DMF, or ethanol that contained 10 g/L of $MgCl_2 \cdot 6H_2O$. We performed gel permeation chromatography, GPC, using Shimadzu LC-10AD high performance liquid chromatograph system with SPD-10A UV-VIS detector, CTO-10A column oven, and chromatopac C-R7A-plus data processor. Shodex GF310HQ and GF510HQ were used as a column. We used DMF that contained 10 g/L of $MgCl_2 \cdot 6H_2O$ as eluate.

Degradation of the Modified Polystyrene by the Treatment with Activated Sludge in Soil

We performed degradation of PSt-*co*-BVP(Cl) by the treatment with activated sludge in soil. A prescribed amount of polymer sample was dissolved in a mixed solvent of ethanol and tetrahydrofuran in a weight ratio 1:1, mixed with the purified soil, and dried to a constant weight under a reduced pressure at room temperature to remove the solvents from the test soil.

We mixed the above-mentioned test soil with a prescribed amount of washed, activated sludge and the artificial sewage. We added the artificial sewage, 0.2 g in COD per g of polymer sample, to assist the biodegradation. The total amount of water and activated sludge were settled to 200 mL/kg and 14.2 g/kg in wet weight, respectively.

We performed the biodegradation by allowing the mixture to stand at room temperature. After a prescribed time, we recovered remained polymer by extraction with a mixture of ethanol and tetrahydrofuran in a weight ratio 1:1 for 48 h using a Soxhlet extraction apparatus. We removed fine soil particles contained in the extractive by centrifugation at 2000 *g* for 20 min. We repeated this centrifugation three times, and removed solvents by evaporation using a rotary evaporator. We added ethyl acetate to the residue to isolate the recovered polymer sample, and dried the isolated polymer to a constant weight under a reduced pressure at room temperature. Prior to these experiments of biological treatment, we performed a series of experiments to ascertain the reliability of the recovery procedure.

DEGRADATION OF THE MODIFIED POLYSTYRENE DURING TREATMENT WITH ACTIVATED SLUDGE IN SOIL

We performed biodegradation of PSt-*co*-BVP(Cl) by the treatment with activated sludge in soil at room temperature. We recovered residual polymer by extraction using ethanol and tetrahydrofuran, and precipitation using ethyl acetate. Thus, low molecular weight oligomers and other organic materials soluble in ethyl acetate were not included in the residual weight.

Figures 1 and 3 show time course of weight reduction during the treatment. Here, residual weight is the total weight of polymeric materials recovered after the treatment. Figures 2 and 4 show time course of $t \ln 2/\ln \{a/(a-x)\}$ during the treatment. Here, t is the treatment time, a is the initial amount of polymer, and x is the amount of polymer disappeared during the treatment time t.

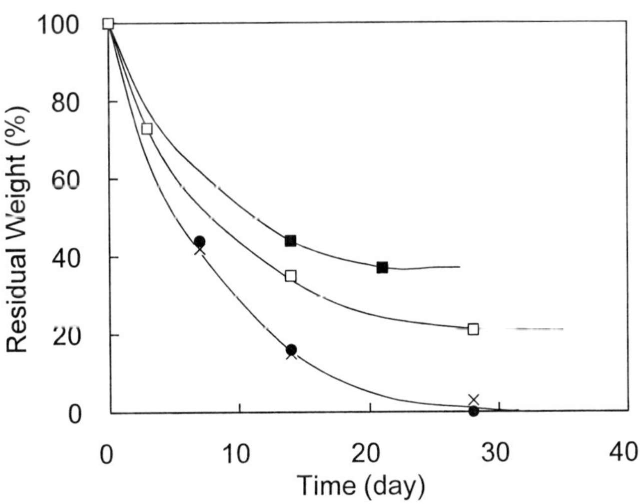

Figure 1. [17]. Time course of weight reduction of PSt-*co*-BVP(Cl) during the treatment with activated sludge in soil at room temperature. The amount of polymer sample used for the treatment is given in parentheses in g/kg. (×) PSt-*co*-BVP(Cl)$_1$ (0.5); (●) PSt-*co*-BVP(Cl)$_1$ (1.0); (□) PSt-*co*-BVP(Cl)$_2$ (0.5); (■) PSt-*co*-BVP(Cl)$_2$ (1.0).

Biodegradation of Pst-*co*-BVP(Cl) In Molar Ratio 1:1

PSt-*co*-BVP(Cl) in molar ratio 1:1, PSt-*co*-BVP$(Cl)_1$, contained 43.1-mol % styrene, 49.1-mol % BVP(Cl) and 7.8-mol % 4-vinylpyridine. Intrinsic viscosity of PSt-*co*-BVP$(Cl)_1$ determined in ethanol that contained 10 g/L of $MgCl_2 \cdot 6H_2O$ at 35°C was 0.26 dL/g.

Cross marks and closed circles in Figure 1 show the time course of weight reduction of PSt-*co*-BVP$(Cl)_1$ obtained by using 0.5 and 1.0 g/kg of the polymer sample, respectively, during the treatment with activated sludge in soil. Time courses obtained under these two conditions were very close.

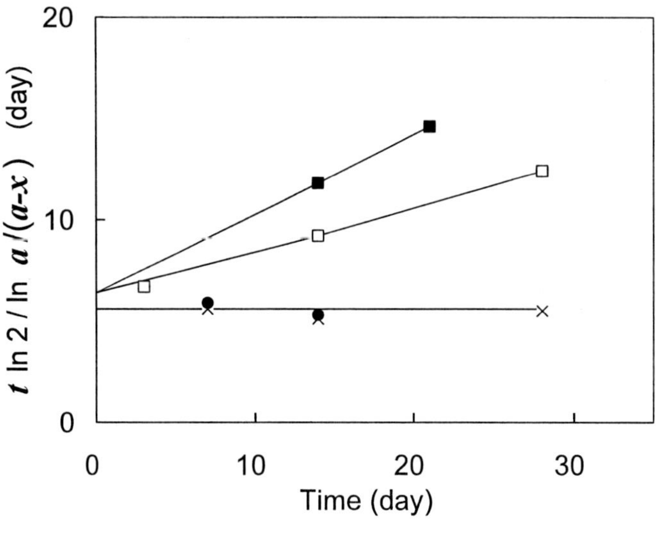

Figure 2. [17]. Time course of $t \ln 2/\ln \{a/(a-x)\}$ during the treatment of PSt-*co*-BVP(Cl) with activated sludge in soil at room temperature. t, treatment time (day); a, initial amount of polymer (g/kg); x, amount of polymer disappeared during the treatment time t (g/kg). The amount of polymer sample used for the treatment is given in parentheses in g/kg. (×) PSt-*co*-BVP$(Cl)_1$ (0.5); (●) PSt-*co*-BVP$(Cl)_1$ (1.0); (□) PSt-*co*-BVP$(Cl)_2$ (0.5); (■) PSt-*co*-BVP$(Cl)_2$ (1.0).

Cross marks and closed circles in Figure 2 show the time course of $t \ln 2/\ln \{a/(a-x)\}$ obtained by using 0.5 and 1.0 g/kg of the polymer sample, respectively, during the biological treatment. Time courses obtained under these two conditions were very close, and did not depend upon the treatment time t.

These experimental results showed that biodegradation of PSt-*co*-BVP(Cl)$_1$ by activated sludge followed first order kinetics under the experimental conditions. The addiction of 1.0 g/kg of the polymer appeared not too much and within the degradation ability of microbes contained in activated sludge used for the biological treatment. Half-life of PSt-*co*-BVP(Cl)$_1$ was 5.6 days.

Biodegradation of PSt-*co*-BVP(Cl) in Molar Ratio 2:1

PSt-co-BVP(Cl) in molar ratio 2:1, PSt-*co*-BVP(Cl)$_2$, contained 63.0-mol % styrene, 35.5-mol % BVP(Cl) and 1.5-mol % 4-vinylpyridine. Molar ratio of styrene to BVP(Cl) was 1.8. The average number of styrene units contained in the oligo-styrene portion was 1.8. The weight average molecular weight of PSt-*co*-BVP(Cl)$_2$ determined by GPC analysis was 116,000.

Open squares and closed squares in Figure 1 show the time course of weight reduction of PSt-*co*-BVP(Cl)$_2$ obtained by using 0.5 and 1.0 g/kg of the polymer sample, respectively during the biological treatment. Making a sharp contrast to the case of PSt-*co*-BVP(Cl)$_1$, weight reduction of PSt-*co*-BVP(Cl)$_2$ was faster when a smaller amount of the polymer sample was used. This difference in weight reduction appears to be derived from the difference of biodegradability between the two polymers. Addition of 0.5 g/kg of PSt-*co*-BVP(Cl)$_2$ appeared too much with respect to the biodegradation ability of microbes contained in activated sludge.

Open squares and closed squares in Figure 2 show the time course of $t \ln 2/\ln \{a/(a\text{-}x)\}$ obtained by using 0.5 and 1.0 g/kg of the polymer sample, respectively, during the biological treatment. The slope was lower when a smaller amount of polymer was used. Biodegradation of PSt-*co*-BVP(Cl)$_2$ did not follow first order kinetics. The addiction of 0.5 g/kg of polymer seemed overloading with respect to the biodegradation ability of microbes contained in activated sludge.

Linear relationship shown in Figure 2 will become independent upon the treatment time under ultimate conditions where the amount of the polymer sample was sufficiently small and was within the biodegradation ability of microbes contained in activated sludge.

On the other hand, as can be seen in Figure 2, the linear relationship shown by open squares coincides with that shown by closed squares at the treatment time zero. Based on these considerations, we estimated half-life of PSt-*co*-BVP(Cl)$_2$ 6.4 days by extension of the linear relationship of Figure 2 to the treatment time zero. This half-life may be reasonable under the ultimate conditions where the amount of the polymer sample was sufficiently small and did not exceed the biodegradation ability of microbes contained in activated sludge.

It is desirable to evaluate the exact half-life by using a sufficiently small amount of the polymer sample, rather than the extension of the linear relation of Figure 2 to the treatment time zero. However, reliable analysis with respect to the residual amount of the polymer requires a minimum amount of the polymer sample. Therefore, we estimated half-life by extension of the linear relationship of Figure 2 to the treatment time zero.

Biodegradation of PSt-*co*-BVP(Cl) in Molar Ratio 3:1

PSt-*co*-BVP(Cl) in molar ratio 3:1, PSt-*co*-$BVP(Cl)_3$, contained 73.5-mol % styrene and 26.5-mol % BVP(Cl). 4-Vinylpyridine was not contained. Molar ratio of styrene to BVP(Cl) was 2.8. The average number of styrene in the oligo-styrene portion was 2.8. The weight average molecular weight of PSt-*co*-$BVP(Cl)_3$ determined by GPC analysis was 105,000.

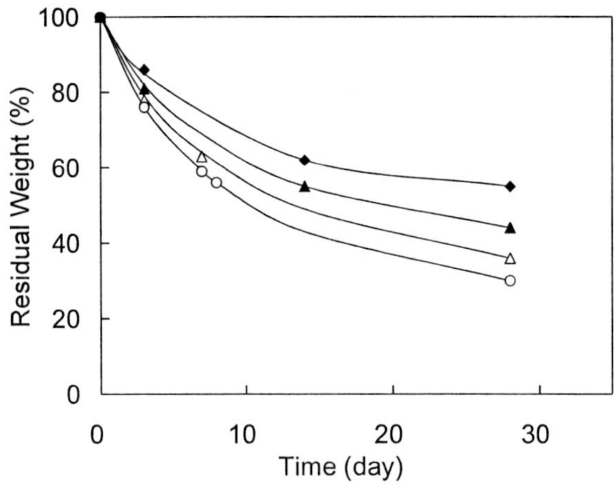

Figure 3. [17]. Time course of weight reduction of PSt-*co*-BVP(Cl) during the treatment with activated sludge in soil at room temperature. The amount of polymer sample used for the treatment was 0.5 g/kg. (○) PSt-*co*-$BVP(Cl)_3$; (△) PSt-*co*-$BVP(Cl)_4$; (▲) PSt-*co*-$BVP(Cl)_5$; (◆) PSt-*co*-$BVP(Cl)_8$.

We performed the biodegradation using 0.5 g/kg of the polymer sample. Open circles in Figures 3 and 4 summarize the results. We estimated half-life 6.9 days by extension of the linear relation of open circles in Figure 4 to time zero. This half-life revealed a striking contribution of BVP(Cl), because no degradation of trimer of styrene by bacterial strain was detected [13].

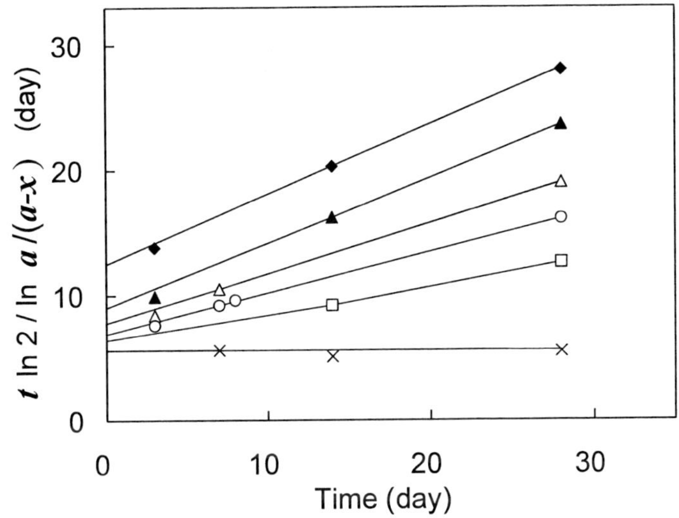

Figure 4. [17]. Time course of t ln 2/ln $\{a/(a-x)\}$ during the treatment with activated sludge in soil at room temperature. t, treatment time (day); a, initial amount of polymer (g/kg); x, amount of polymer disappeared during the treatment time t (g/kg). The amount of polymer sample used for the treatment was 0.5 g/kg. (×) PSt-*co*-BVP(Cl)$_1$; (□) PSt-*co*-BVP(Cl)$_2$; (○) PSt-*co*-BVP(Cl)$_3$; (△) PSt-*co*-BVP(Cl)$_4$; (▲) PSt-*co*-BVP(Cl)$_5$; (◆) PSt-*co*-BVP(Cl)$_8$.

Biodegradation of PSt-*co*-BVP(Cl) in Molar Ratio 4:1

PSt-*co*-BVP(Cl) in molar ratio 4:1, PSt-*co*-BVP(Cl)$_4$, contained 80.2-mol % styrene, 19.4-mol % BVP(Cl) and 0.4-mol % 4-vinylpyridine. Molar ratio of styrene to BVP(Cl) was 4.2. The average number of styrene units in the oligo-styrene portion was 4.2. The weight average molecular weight of PSt-*co*-BVP(Cl)$_4$ determined by GPC analysis was 79,000.

We performed the biodegradation using 0.5 g/kg of a polymer sample. Open triangles in Figures 3 and 4 summarize the results. We estimated half-life 7.8 days by the extension of the linear relation of open triangles in Figure 4 to the treatment time zero.

Biodegradation of PSt-*co*-BVP(Cl) in Molar Ratio 5:1

PSt-*co*-BVP(Cl) in molar ratio 5:1, PSt-*co*-BVP$(Cl)_5$, contained 84.0-mol % styrene, 15.7 mol% BVP(Cl) and 0.4-mol % 4-vinylpyridine. Molar ratio of styrene to BVP(Cl) was 5.4. The average number of styrene units in the oligo-styrene portion was 5.4. The weight average molecular weight of PSt-*co*-BVP$(Cl)_5$ determined by GPC analysis was 73,900.

We performed the biodegradation using 0.5 g/kg of polymer sample. Closed triangles in Figures 3 and 4 summarize the results. We estimated half-life 9.0 days by the extension of the linear relation of closed triangle in Figure 4 to the treatment time zero.

Biodegradation of PSt-*co*-BVP(Cl) in Molar Ratio 8:1

PSt-*co*-BVP(Cl) in molar ratio 8:1, PSt-*co*-BVP$(Cl)_8$, contained 88.0-mol % styrene, 10.6-mol % BVP(Cl) and 1.5-mol % 4-vinylpyridine. Molar ratio of styrene to BVP(Cl) was 8.3. The average number of styrene units in the oligo-styrene portion was 8.3. The weight average molecular weight of PSt-*co*-BVP$(Cl)_8$ determined by GPC analysis was 264,000.

We performed the biodegradation using 0.5 g/kg of polymer sample. Closed rhombuses in Figures 3 and 4 summarize the results. We estimated half-life 12.5 days by the extension of the linear relation of closed rhombuses in Figure 4 to the treatment time zero.

Mode of Chain Scission in Biodegradation of the Modified Polystyrene

The mode of chain scission was investigated using PSt-*co*-BVP$(Cl)_3$. Since the recovered polymer sample did not exhibit perceivable change of elementary composition, we estimated the rate of chain shortening of the recovered polymer based on the rate of reduction of molecular weight.

In the case where first scission occurs at the center of the polymer chain, the molecular weight of the recovered polymer may dramatically reduce to half, but the decrease of gravimetric weight may be negligible at this stage. Similarly, when random scission at the inside of the polymer chain predominated, the reduction of molecular weight of the recovered polymer may predominate over the decrease of gravimetric weight.

On the other hand, when the first scission occurs at the end of the polymer chain, the decrease of gravimetric weight as well as the reduction of molecular weight of the recovered polymer may be negligible at this stage. Similarly, in the case of uniformly successive scission from the end of polymer chain, the reduction of molecular weight of the recovered polymer and decrease of gravimetric weight may occur in parallel.

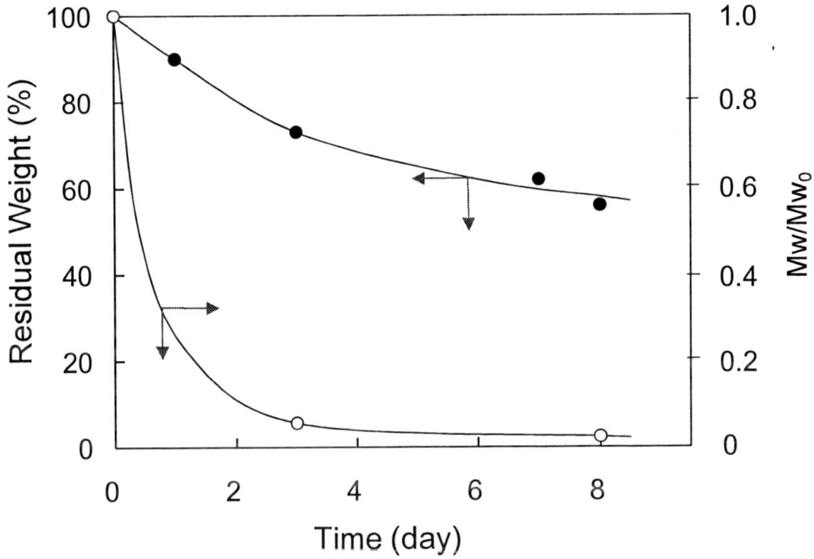

Figure 5. [17]. Time course of residual weight (●) and rate of reduction of molecular weight of recovered polymer during the treatment of PSt-*co*-BVP(Cl)$_3$ (○) during the treatment with activated sludge in soil at room temperature. Mw is the molecular weight of recovered polymer, and Mw_0 is the molecular weight of the polymer before the biological treatment. The amount of polymer sample used for the treatment was 0.5 g/kg.

Time course of residual weight is shown by closed circles in Figure 5. The amount of the polymer sample was 0.5 g/kg. Open circles in Figure 5 show the time course of rate of reduction of molecular weight of the recovered polymer. Obviously, the reduction of molecular weight predominated over the decrease of gravimetric weight. This result clearly indicates that cleavage predominated at the BVP(Cl) portion over the trimer portion. Biodegradation of trimer of styrene appeared to be conspicuously more difficult than BVP(Cl), although dimer of styrene is biodegradable [13].

INFLUENCE OF SIZE OF OLIGOMER PORTION ON BIODEGRADABILITY OF THE MODIFIED POLYSTYRENE

Figure 6 shows the relation between half-life of PSt-*co*-BVP(Cl) and the average number of styrene units in the oligo-styrene portion. Except for PSt-*co*-BVP(Cl)$_1$, we estimated half-lives by the extension of the linear relation in Figures 2 and 4 to the treatment time zero. These half-lives are significant under ultimate conditions where the amount of polymer samples was sufficiently small and did not exceed the degradation ability of microbes contained in the activated sludge.

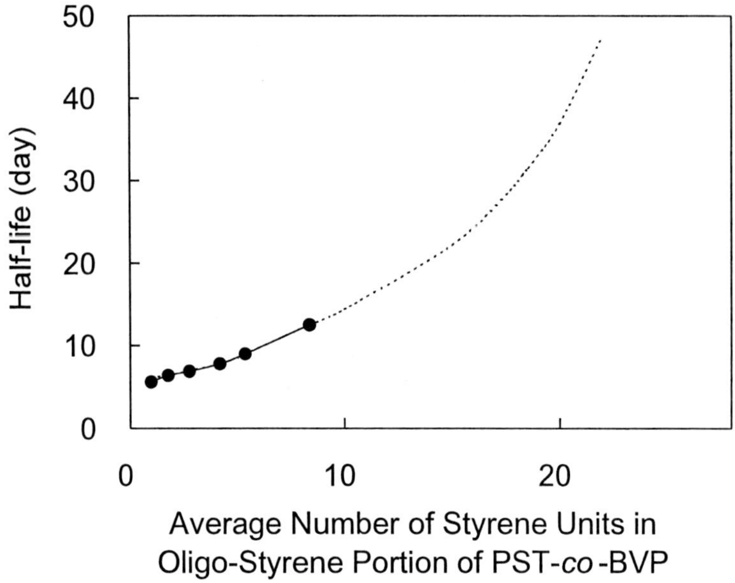

Figure 6. [17]. Relation between half-life and average number of styrene units in the oligo-styrene portion of PSt-*co*-BVP(Cl).

Figure 6 suggests that half-life could be in the range of 30-40 days when PSt-*co*-BVP(Cl) contained 5-mol % BVP(Cl), *i.e.,* oligo-styrene portion contained 20 styrene units. The insertion of 5-mol % BVP(Cl) would be sufficient to make polystyrene substantially biodegradable, if we do not expect extraordinarily rapid biodegradation in the natural environment, although such biodegradation could be realized when a very low amount of polymer sample was emitted.

Intermediate Materials Produced During the Biodegradation

After the treatment of PSt-*co*-BVP(Cl) with activated sludge in soil, we recovered residual polymer by extraction using a mixture of ethanol and tetrahydrofuran in a weight ratio 1:1, as described in the experimental section. After the removal of soil particles from the extracts by centrifugation, we removed solvents by evaporation using a rotary evaporator, and added ethyl acetate to the residue to precipitate and isolate the recovered polymer. However, we found intermediate materials in the layer of ethyl acetate, and tried to isolate and identify them. Time course of total amount of the intermediate products is shown in Figure 7 in the cases of PSt-*co*-BVP$(Cl)_2$ (closed squares), PSt-*co*-BVP$(Cl)_3$ (closed circles), and PSt-*co*-BVP$(Cl)_4$ (closed triangles) together with time course of residual weight of PSt-*co*-BVP$(Cl)_2$ (open squares), PSt-*co*-BVP$(Cl)_3$ (open circles), and PSt-*co*-BVP$(Cl)_4$ (open triangles).

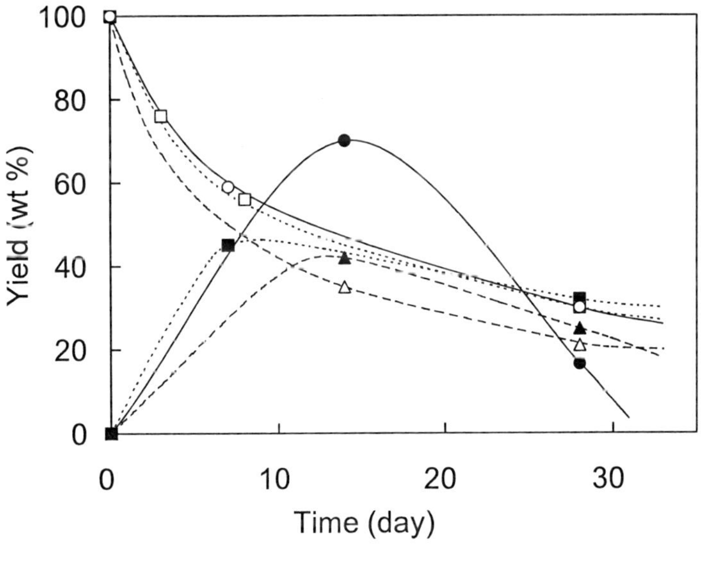

Figure 7. [17]. Time course of the yields of residual polymer (□, ○, △) and low molecular weight products that were soluble in ethyl acetate (■, ●, ▲) during the treatment of PSt-*co*-BVP$(Cl)_2$ (□ and ■), PSt-*co*-BVP$(Cl)_3$ (○ and ●), and PSt-*co*-BVP$(Cl)_4$ (△ and ▲), respectively, with activated sludge in soil. The amount of polymer sample used for the treatment was 0.5 g/kg.

We made every effort to identify the recovered intermediates but could not accomplish. Ultraviolet spectrum of the recovery showed a maximum absorbance at 275 nm. Column chromatography using silica gel as the stationary phase and ethyl acetate as the eluate revealed that the recovery was a complicated mixture of several components. Unfortunately, we could not separate and identify these individual compounds definitively, although elementary analyses, FTIR spectrum, ^{1}H and ^{13}C NMR spectra suggested the presence of carboxyl group in the intermediate products, in addition to the presence of phenyl group and BVP(Cl).

In the biological treatment of copolymers of BVP(Cl) with methyl methacrylate [24] and acrylonitrile [26] with activated sludge in soil, we did not obtain such intermediate materials. Recovery of intermediate products during the biological treatment of PSt-*co*-BVP(Cl) could reflect the extraordinarily difficult biodegradation of polystyrene and oligo-styrene.

A Variety of Noticeable Utilities of the Modified Polystyrene

We found a variety of noticeable utilities of PSt-*co*-BVP(Cl) in the field of water purification and biotechnology due to the strong affinity of BVP(Cl) with microbes and viruses.

Removal of Microbes from Water by Filtration

Non-woven cloth coated with 32 mg/g of PSt-*co*-$BVP(Cl)_1$ was effective for removing microbes from water [27]. In removal of *E. coli* by filtration through ten sheets of the coated non-woven cloth, the rate of removal was 99.99% at a filtration rate of 2.6 cm/h. The rate of removal remained at 99% even at a high filtration rate of 300 cm/h and a low influent concentration of the bacterial cells such as 10^3 cells/mL. The rate of removal tended to increase with a decrease in the influent bacterial concentration. Seven other bacteria and two yeasts were effectively removed by filtration through the coated non-woven cloth. The filtration was also effective in removing spores of fungi from water, but was not very effective in removing bacteriophage T4 from aqueous solution.

Removal of Viruses from Water by Filtration

Removal of viruses by filtration is extremely difficult because viruses are very small. However, PSt-*co*-BVP(Cl) is useful for effective removal of virus by filtration. Cross-linked PBVP(Br) strongly adsorbs bacteriophage T4 [28] and pathogenic human viruses [29].

Removal of virus by filtration using micro-porous cellulose and cellulose acetate membranes was reported [30-32], but efficiency of the virus removal was remarkably improved when the membranes were coated with PSt-*co*-BVP(Cl) [33]. Conventional ultra filtration using 1-3 sheets of 145-μm-thick cellulose nitrate membrane coated with 1.7 mg/g of PSt-*co*-BVP(Cl) in molar ratio 5:2, pore size was 0.45 μm, showed 99.4-99.998% removal (2.2-4.7 $\log_{10}$-unit reduction in the concentration) of bacteriophage T4. On the other hand, control experiments using the corresponding non-coated membranes showed in the range of 91-96% removal (1.0-1.4 $\log_{10}$-unit reduction in the concentration) of bacteriophage T4.

Bioreactor with Use of Captured Bacterial Cells

We immobilized cells of *E. coli* having aspartase activity on the surface of non-woven cloth coated with 10 mg/g PSt-co-BVP(Cl) in molar ratio 5:2 [34]. Continuous operation of a fixed-bed column reactor containing 21.7 g/L of the immobilized cells produced L-aspartic acid in 95% yield from ammonium fumarate. Immobilization on the coated non-woven cloth insignificantly affected optional pH of the biochemical reaction. L-aspartic acid was obtained in 77% yield after 160 days of continuous operation, and the initial yield was estimated to require about 500 days for halving.

We used cells of *Arthrobacter simplex* having 3-ketosteroid-Δ^1-dehydromonogenase activity immobilized by the capture on the surface of non-woven cloth coated with 6.5 g/m^2 PSt-*co*-BVP(Br) in molar ratio 3:1 for dehydrogenation of cortisol to prepare prednisolone [35]. Using a fixed-bed column reactor, we obtained prednisolone in about 90% yield during 6 days.

Rapid Determination of Concentration of Viable Bacteria

We prepared an electrochemical sensor that rapidly determined concentrations of viable cells of aerobic and facultatively anaerobic bacteria

[36, 37]. It consisted of a flow cell that contained an oxygen electrode and non-woven cloth coated with PSt-*co*-BVP(Cl) in molar ratio 5:2. When a sample suspension of microbes was passed through the electrochemical sensor at a constant flow rate, the microbes were captured on the surface of coated non-woven cloth in the living state. Consumption of oxygen by the captured living microbes appeared as a decrease in the electric current. Rate of decrease of the electric current was proportional to the concentration of viable microbes. Each measurement required only about 10 minutes. The minimum detectable concentration of living microbes expressed as turbidity at 610 nm was 0.005-0.023. In the case of *E. coli*, this minimum detectable concentration corresponded to the conventional colony count 3.7×10^5 cells/mL.

Coagulation and Sedimentation of Microbes in Water

Sawdust coated with 1 wt % PSt-*co*-BVP$(Cl)_1$ produced coagulation and sedimentation of bacteria [38]. When the coated sawdust was mixed with suspensions of *E. coli* and allowed to stand, the coated sawdust precipitated, and the bacterial population in the supernatant layer reduced to 1/1,000,000 or less during the procedure under appropriate conditions. The bacterial cells formed flocks and accumulated over the precipitated sawdust. This result suggests coagulation of bacterial cells in addition to the capture of bacterial cells on the surface.

Control of Soil-borne Plant Diseases

Soil-borne plant diseases can be controlled by the addition of sawdust coated with 1 wt % PSt-*co*-BVP$(Cl)_1$. Tomato bacterial wilt caused by *Ralstonia solanacearum* was controlled by the addition of the coated sawdust to soil prior to transplantation of seedlings [39]. This effected 87% reduction in appearance and 89% reduction in the index of symptoms under appropriate conditions. The disease control was explained in terms of the reduction of infectious contact between roots of tomato and cells of *R. solanacearum* due to coagulation-like interaction between microbial cells and the capture of the bacterial cells by the coated sawdust.

REFERENCES

[1] Byrom, D. (Ed.), *Biomaterials*, MacMillan, London, 1991.
[2] Doi, Y. (Ed.), *Handbook of Biodegradable Plastics*, NTS Inc., Tokyo, 1995.
[3] Dumitriu, S., *Polymeric Biomaterials*, CRC Press, New York, 2001.
[4] Steinbuchel, A. (Ed.), *Biopolymers: Cumulative Index*, Wiley-Vch, Weinheim, 2003.
[5] Smith, R. (Ed.), *Biodegradable Polymers for Industrial Applications*, Woodhead Publ., Cambridge, 2005.
[6] Mallapragada, S. K.; Narasimhan, B. (Eds.), *Handbook of Biodegradable Polymeric Materials and Their Applications,* American Scientific Publ., California, 2005.
[7] Bastioli, C. (Ed.), *Handbook of Biodegradable Polymers*, Rapra Technol., London, 2005.
[8] Shimao, M.; Ninomiya, K.; Kuno, O.; Kato, N.; Sakazawa, C., *Appl. Environ. Microbiol.* 51 (1986) 268-275.
[9] Albertson, A. –C.; Banhidi, Z. G., *J. Appl. Polym. Sci.* 25 (1980) 1655-1671.
[10] Tsuchii, A.; Suzuki, T.; Fukuoka, S., *Agric. Biol. Chem.* 48 (1984) 621-625.
[11] Tsuchii, A.; Suzuki, T.; Takahara, Y., *Agric. Biol. Chem.* 43 (1979) 2441-2446.
[12] Hosoya, H.; Miyazaki, N.; Sugisaki, Y.; Takanashi, E.; Tsurufuji, M.; Yamasaki, M.; Tamura, G., *Agric. Biol. Chem.* 42 (1978) 1545-1552.
[13] Tsuchii, A.; Suzuki, T.; Takahara, Y., *Agric. Biol. Chem.* 41 (1977) 2417-2421.
[14] Matsumura, S.; Ii, S.; Shigeno, H.; Tanaka, T.; Okuda, F.; Shimura, Y.; Toshima, K., *Makromol. Chem.* 194 (1993) 3237-3246.
[15] Kawabata, N.; Hayashi, T.; Matsumoto, T., *Appl. Environ. Microbiol.* 46 (1983) 203-210.
[16] Kawabata, N.; Utihori, D.; Fukuda, S.; Funahashi, H., *J. Appl. Polym. Sci.* 51 (1994) 33-42.
[17] Kawabata, N.; Kakimoto, C.; Ikeuchi, H., *J. Appl. Polym. Sci.* 100 (2006) 554-559.
[18] Kawabata, N.; Fuse, T., *J. Appl. Polym. Sci.* 100 (2006) 1618-1623.
[19] Maeda, Y., *Hakko Kogaku Kaishi* 57 (1979) 114-124.
[20] Horazawa, I. (Ed.), *Bio-Film Process*, Shikosha, Tokyo, 1982.
[21] Rittmann, B. E., *Environ. Sci. Technol.* 21 (1987) 128-136.

[22] Kawabata, N.; Nishiguchi, M., *Appl. Environ. Microbiol.* 54 (1988) 2532-2535.

[23] Kawabata, N.; Hayashi, T.; Nishikawa, M., *Bull. Chem. Soc. Jpn.* 59 (1986) 2861-2863.

[24] Kawabata, N.; Ueno, Y.; Torii, K.; Matsumoto, T., *Agric. Biol. Chem.* 51 (1987) 1085-1090.

[25] Rosenberg, M.; Gutnick, D.; Rosenberg, E., *FEMS Microbil. Lett.* 9 (1980) 29-33.

[26] Katayama, T.; Takai, K.; Okabe, A.; Hayashi, H.; Kanemasa, Y., *Nippon Saikingaku Zasshi* 30 (1975) 121.

[27] Kawabata, N.; Inoue, T.; Tomita, H., *Epidemiol. Infect.* 108 (1992) 123-134.

[28] Kawabata, N.; Hashizume, T.; Matsumoto, T., *Agric. Biol. Chem.* 50 (1986) 1551-1555.

[29] Kawabata, N.; Yamazaki, K.; Otake, T.; Oishi, I.; Minekawa, Y., *Epidemiol. Infect.* 105 (1990) 633-642.

[30] Manabe, S.; Kamata, Y.; Iijima, H.; Kamide, K., *Polymer J.* 19 (1987) 391-404.

[31] Manabe, S.; Tsurumi, T.; Ishikawa, G.; Satani, M.; Yamasaki, T.; Yamaguchi, K.; Kobayashi, S.; Yamamoto, N., *Membrane* 14 (1989) 77-83.

[32] Sekiguchi, S.; Ito, K.; Kobayashi, M.; Ikeda, H.; Tsurumi, T.; Ishikawa, G..; Manabe, S.; Satani, M.; Yanashiki, T., *Membrane* 14 (1989) 253-261.

[33] Kawabata, N.; Fujita, I.; Inoue, T., *J. Appl. Polym. Sci.* 60 (1996) 911-917.

[34] Kawabata, N.; Hatanaka, H.; Odaka, H., *J. Ferment. Bioeng.* 79 (1995) 317-322.

[35] Kawabata, N.; Nakagawa, K., *J. Ferment. Bioeng*, 71 (1991) 19-23.

[36] Kawabata, N.; Teramoto, K.; Ueda, T., *J. Microbilol. Methods* 15 (1992) 101-111.

[37] Kawabata, N.; Teramoto, K., *Sensors & Aquators B* 13-14 (1993) 309-311.

[38] Kawabata, N.; Tanabe, E., *React. Funct.Polym.* 65 (2005) 293-299.

[39] Kawabata, N.; Kishimoto, H.; Abe, T.; Ikawa, T.; Yamanaka, K.; Ikeuchi, H.; Kakimoto, C., *Biosci. Biotechnol. Biochem.* 69 (2005) 326-333.

In: Polystyrene: Properties, Performance… ISBN: 978-1-61209-117-4
Editor: James E. Gray

Chapter 3

STUDY OF THE ADHESION MECHANISMS OF PARTICLES ON MODIFIED POLYPROPYLENE: INFLUENCE OF SURFACE PARAMETERS

J. Delattre[1, 2], *N. Anjum*[1, 2], *A. Guinault*[3] *and A.M. Riquet*[1, 2]
[1] INRA, UMR 1145, Massy, France
[2] AgroParisTech, UMR 1145, Massy, France
[3] CNAM MPI, Paris cédex, France

ABSTRACT

The present work should be considered in the context of the health safety and hygiene of materials.

In several areas of application of plastic materials (agro-food industry, medical devices or even pharmaceuticals) it is necessary to understand how a surface reacts in contact with the surrounding environment (food, living beings, etc.) so that surfaces can be developed that meet their requirements. The aim of this work was therefore to understand the role of surface parameters implicated in adhesion phenomena. The use of spherical particles (i.e. polystyrene or silica beads) as a model was designed to enable adhesion tests during which their surface properties would not vary.

The first challenge of this work was to modify the surface of polypropylene (PP) by radio-grafting with hydrophilic functional molecules such as N,N-dimethylacrylamide (DMA) and [2-methacryloy loxy)ethyl]trimethylammonium chloride, which is a quaternary ammonium salt (QAS). These monomers were used alone or in well defined mixtures. The aim of grafting was to modulate the surface parameters of PP such as wettability, roughness and surface charge.

Determination of the degree of grafting by weighing combined with infrared (FTIR-ATR) and X-Ray photoelectron spectroscopy (XPS) analysis of the surface confirmed the grafting of the molecules. Analysis of the physicochemical characteristics of the modified surfaces showed that grafting had changed these surface parameters as a function of the grafting conditions (concentration of grafting solution, type of molecule).

The second part of this work was designed to understand the mechanisms involved in adhesion phenomena. Particles adhering to the modified samples were studied according to the surface parameters.

Adhesion tests confirmed the strong influence of substrate type (mainly hydrophilicity and roughness) and to a lesser extent underlined the role of electrostatic interactions.

INTRODUCTION

The microbiological quality of foods is a major public health and economic challenge. Indeed, the contamination of products by pathogenic micro-organisms can cause food poisoning. In recent years, new infections have appeared, others have seen a resurgence and cases of resistance have developed, not to mention new modes of transmission related to the use of novel processes. Current trends in consumption (i.e. more fresh products, more ready-to-eat meals), new technologies, the globalisation of markets and the adaptability of micro-organisms are all factors which today require improvements to the control of food safety.

The control of food safety implies the safety of the microbiological contamination of surfaces that may be in contact with food products. Numerous studies have shown that the contamination of products by pathogenic organisms and/or by alteration generally results from their adhesion to surfaces. This adhesion is the first step in the formation of biofilms 1, 2], a permanent source of biocontamination in industry. Among these undesirable micro-organisms, *Listeria monocytogenes*, an ever-present bacteria involved in food poisoning, occupies a preponderant place at the same level as *Salmonella* Typhimurium or even *Escherichia coli.* These gram-positive or gram-negative bacilli can survive and develop under particularly difficult conditions (i.e. low temperatures, acid pH levels, highly saline environments) [3] and adhere to a wide variety of surfaces [4, 5]. They can be found throughout the food chain, particularly in rinsing water and on equipment surfaces [6]. It should be remembered that the formation of biofilms gives rise to increased production costs (impaired equipment performance, higher concentrations of washing products and longer processing times, etc.), losses associated with the premature deterioration of finished products and, of course, costs related to the recall of products containing pathogenic organisms and the consequences of any resulting cases of food poisoning. It is clear today that biofilms constitute permanent reservoirs of contamination that are extremely difficult to eradicate. It is therefore necessary to develop strategies which will ensure that surfaces in contact with foods will not serve as a vehicle for pathogen transmission. One way is to control the physicochemical interactions that occur between the bacteria and the substratum surface.

Under the terms of both traditional and extended DLVO theories, the colonisation of solid surfaces is governed by the surface properties of the two bodies. Adhesion phenomena in an aqueous medium can be described as an equilibrium between Lifshitz-van der Waals interactions, Lewis acid-base interactions, electrostatic interactions and their reduction as a function of the distance between the two bodies [7, 8]:

- •Lifshitz-van der Waals interactions depend on the apolar component of free surface energy, and are mainly attractive.
- •Lewis acid-base interactions are dependent on the polar electron donor and acceptor components of free surface energy and can attract or repel [9].
- •Electrostatic interactions are due to overlapping of the double ionic layer associated with the strength of the surface charges placed in contact [10]. This interaction is repulsive if the surface charges have the same sign or, on the contrary, attracts, declining exponentially as a function of the distance separating the two bodies.

Surface charges and free surface energies are responsible for long and short distance interactions. Long distance interactions are described as an equilibrium between Lifshitz-van der Waals interactions and electrostatic interactions, while short distance interactions mainly involve Lewis acid-base interactions.

In order to limit surface contamination by micro-organisms, the approach adapted during the present study consisted in modifying the surface properties of polymeric material (hydrophily, energy characteristics and surface charge) by grafting hydrophilic functional molecules such as acrylic acid (AAc) N,N-dimethylacrylamide (DMA) and [2-methacryloyloxy)ethyl] trimethyl-ammonium chloride, which is a quaternary ammonium salt (QAS). Quaternary ammonium salts (QASs) are synthetic organic chemicals that are widely used in a variety of areas such as environmental disinfection, cosmetics, ophthalmic solutions, pharmaceutical preparations and to control fungal infections [11]. QASs exert an antimicrobial effect by damaging the cytoplasmic membrane [12]. However, this effect is due to the presence of alkyl chains and their surfactant properties [13]. QASs have stronger antibacterial effects on gram positive bacteria than on gram negative bacteria because they have an extra protective membrane which inhibits the diffusion of numerous chemical products [14]. In the present study, a QAS was used to modulate the surface energy properties of the substrate and also to control bacterial adhesion. Its usefulness lies in its high electron acceptor capacity.

The aim of this work was therefore to understand the mechanisms involved in adhesion phenomena. Adhesion tests were performed using silica and polystyrene beads with surface physicochemical properties similar to those of some *Listeria monocytogenes* strains [15]. Indeed, these beads constitute a less complex model to understand adhesion phenomena insofar as their surface properties do not vary during contact with a surface, unlike micro-organisms.

During the first part of the study, polypropylene (a material that plays a dominant role in the agro-food industry) was modified by the grafting of hydrophilic monomers (anionic, cationic and neutral) using a pre-irradiation method. The polymer sheets were activated by electron beam irradiation in air under ambient conditions so that radicals were generated in the polymer matrix; these were subsequently transformed into peroxides and initiated the grafting process. The surface properties (atomic composition, wettability, roughness) of the ungrafted and grafted sheets prepared under different grafting conditions were investigated.

Secondly, we tried to understand the mechanisms involved in the adhesion of silica and polystyrene (PS) beads to the surface of modified polypropylene surfaces. This study was performed by measuring their energy characteristics (Lifshitz-van der Waals and polar components). The zeta potentials of ungrafted and grafted PP were also determined by streaming potential measurements to obtain the surface charge [16, 17, 18] of the two bodies present.

EXPERIMENTAL

Materials and Modification of PP Surfaces

Materials

Polypropylene (PP) sheets of 3 x 2 cm^2 were used. These samples were shaped by injection. Two kinds of PP with a different initial surface roughness are studied. The first was a PP named PP micrometer (PP_M) which is in common commercial use (PPH7060, Total). The second had a smoother surface and was referred to PP nanometer (PP_N) (Exxon Mobil). These samples, respectively 1 and 2 mm thick, were used as substrates during surface modification procedures.

Prior to irradiation, the PP samples were washed with a 2% RBS-35 solution in water at 40°C for 10 minutes, followed by five rinses with tap water at 40°C and five rinses in distilled water at room temperature. The samples were dried overnight in an oven at 40°C.

Modifying the surface of inert plastic materials (i.e. without reactive functional groups) imposes a first step known as "radicalar activation" in order to generate free radicals (or highly reactive sites) on the plastic surface. This radicalar activation of a surface can be achieved using ionizing radiation such as electron beams. In a second step, the free radicals bond covalently to the hydrophilic monomers

Irradiation

The PP sheets were activated by electron beam irradiation (low energy electron accelerator LAB-UNIT – Energy Science). The energy of the electron beam was 165 keV, with a beam current of 5 mA and a speed of 18 $f.min^{-1}$. Irradiation was performed in air at a dose of 100 kGy.

Grafting Reaction

The grafting reaction was performed in a closed reactor [19]. The monomer solution was diluted in distilled water. Mohr's salt was added (0.25%) to prevent the homopolymerization of the monomer solution [20, 21]. The reactor was placed in an oven under argon bubbling to obtain an inert atmosphere. The monomer solution was heated to 60°C. The irradiated PP sample was then immersed in the solution in the reactor. After the desired period, the grafted PP sample was removed and placed in distilled water in an ultrasonic bath at 40°C for 10 minutes to remove all residual traces of the monomer. The grafted PP sample was then dried an overnight in an oven at 40°C.

The PP surface was modified by the grafting of hydrophilic functional molecules such as acrylic acid (AA), N,N-diméthylacrylamide (DMA) and [2-(Methacryloyloxy)ethyl]-trimethylammonium chloride, which is a chloride ammonium salt (QAS) (Aldrich). The advantage of using QAS is linked to its high electron acceptor nature.

The monomers were grafted alone or in a defined mixture. [DMA/QAS] was grafted at the following ratios: [10 : 0], [20 : 0], [40 : 0], [20 : 40] and [0 : 40], respectively denoted D1Q0, D2Q0, D4Q0, D2Q4 and D0Q4. Similarly, [AA : QAS] was grafted at the following ratios:

[20 : 0] and [20 : 40], referred to as A2Q0 and A2Q4.

Following the surface modification of PP, the degree of grafting was calculated according to the expression (1):

$$\text{Degree of grafting (\%)} = \frac{W_g - W_0}{W_0} \times 100$$

where W_0 and W_g are the weights of ungrafted and grafted PP sheets, respectively.

Characterization of Surface Properties

Infrared Spectroscopy

Analysis of the modified surfaces was ensured using infrared spectroscopy Fourier transform attenuated total reflection mode, or FTIR-ATR (Bomem). This technique reflects the nature of the chemical functions grafted on the extreme surface. The sample was compressed against a ZnSe crystal at 45°. Twenty scans were performed to determine the spectrum of each analyzed sample.

X-Ray Photoelectron Spectroscopy (XPS)

XPS analysis can identify the atoms present on a surface and define the type of bonds involved in the structure. XPS measurements were performed using an ESCALAB 250 equipped with a non-chromatised Al Ka X-Ray source from Electron Corporation. The pass energy for the general spectrum was 100 eV. The analysis was performed under 6×10^{-9} mbar UHV, while 8×10^{-8} mbar UHV was used to determine erosion. An argon flow was added in the ion gun. During analysis, X-rays reached the sample surface at an angle of 45°. However, the angle of the detection column to the sample surface was 90°. The spectra of C_{1s}, O_{1s} and N_{1s} were calibrated for binding energies at 285, 531 and 401 eV, respectively.

Contact Angle Measurement and Energy Characteristics of Modified Surfaces

Contact angle measurements can determine the hydrophilic/hydrophobic nature of surfaces under study, as well as their energy characteristics.

Contact angles were measured with a G40 Goniometer (Krüss, France) at room temperature using the sessile drop method with three pure liquids on a known surface: high purity water (Millipore milliQ), diiodomethane (Sigma, France) and ethylene glycol (Sigma, France).

Contact angles with droplets at rest are referred to below as equilibrium contact angles.

The energy properties of the samples were obtained using the Young–van Oss equation (2): [22]

$$(1+\cos\theta)\gamma_L = 2\left[\left(\gamma_S^{LW}\gamma_L^{LW}\right)^{1/2} + \left(\gamma_S^{+}\gamma_L^{-}\right)^{1/2} + \left(\gamma_S^{-}\gamma_L^{+}\right)^{1/2}\right]$$

where γ_L is the total surface tension, γ^{LW}, γ^{+} and γ^{-} the Lifshitz-van der Waals, electron-acceptor and electron-donor components of surface tension, respectively, θ is the contact angle and the subscripts S and L relate to the solid sample and liquid, respectively.

The polar component γ^{AB} was calculated according to the expression (3):

$$\gamma_{AB} = 2\sqrt{\gamma_S^{+}\gamma_S^{-}}$$

Zeta Potential Measurement

The zeta potential measurement defines the charge of the surface under analysis. The zeta potentials of the PP sheets were determined by streaming potential measurements in NaCl 1.5×10^{-3} M using a zetameter (SurPASS, AntonPaar). The pH was adjusted within the range of 2-7 by adding KOH or HNO_3.

The streaming potential ΔE is dependent on both the surface charge in the diffuse layer and electrolyte properties, i.e. conductivity K_{sol}, viscosity η and dielectric constant D. For a flat surface such as PP sheets, the streaming potential (ΔE) is related to the zeta potential, so in this case, by knowing ΔE, the zeta potential (ζ) could be calculated using the following equation (4):

$$\zeta = \frac{4\pi\eta}{D}\left(K_{\mathrm{sol}} + \frac{L_{\mathrm{surf}}}{b}\right)\frac{\Delta E}{\Delta P}$$

where L_{surf} represents specific conductance at the surface and b the half distance between the two samples. Under our experimental conditions, the value of b was 0.05 mm.

Roughness

Surface roughness was determined using a method of non-contact with the analyzed surface (Altisurf® 500, Altimet). A probe equipped with a high chromatic aberration lens breaks a light beam with lengths of monochromatic waves over a distance of 300 microns. Depending on the topography of the surface analyzed, certain wavelengths will be reflected. In this case, they were interpreted in terms of altitude and were able to determine parameter Ra (arithmetic roughness deviations from the mean), expressed in microns.

Scanning Electron Microscopy (Sem)

Surface topography was observed using SEM (EVO MA 10, Zeiss). The surfaces were observed at two magnifications: x300 and x1000.

Adhesion of Particles

The adhesion of particles on PP surfaces enables a count of the number of adhered particles and comparison of the results as a function of surface parameters.

The adhesion of 2 μm-diameter silica beads (Bangs Laboratories) and PS beads (Sigma) was studied. The PP sheets were immersed in a solution of beads at a concentration of $5\times10^{-3}\%$ diluted in a NaCl solution at 1.5×10^{-3} M. For the adhesion of silica and PS beads, samples are immersed for one and three hours, respectively. This period was determined from the sedimentation time of each type of beads. After the adhesion test, the samples were rinsed gently with ultra-pure water to remove non-adherent beads and then dried overnight in an oven at 40°C.

Observations and Interpretation

The surfaces were examined under an optical microscope (BX51, Olympus) though a 40× objective. The microscope was connected to a camera (SC30, Olympus) and a computer for image acquisition and storage.

Ten photos were taken of each sample at different locations. For the rougher surfaces, several photos were taken at each location by varying the depth of field.

Images were analyzed to determine the number of adherent beads using a image analysis software (Image Tool).

RESULTS AND DISCUSSION

Modification of PP Surfaces

The results presented in this section refer to those obtained after the surface modification of micrometric PP (PP_M). Because of this, we have identified the most relevant modifications for further study, and modified the surfaces of nanometric PP (PP_N) at the selected proportions of monomers.

Micrometric PP (PP_M)

The curves shown in Figure 1a,b reflect variations in the degree of grafting for N, N dimethylacrylamide DMA (1a) and acrylic acid AAc (1b) as a function of time and at different monomer concentrations. In the case of pure DMA and diluted monomers (10-80%), the degree of grafting increased with time, even after 40 minutes. With pure AAc, it increased with time up to 20 min and then tended to reach equilibrium. The point of particular interest arising from these results was that grafting increased in line with the percentages of AAc up to 80%, but beyond that level tended to decline to a significant extent (Figure 2). These results, already observed elsewhere [23, 24] indicated that the AAc grafting process was under the cumulative influence of at least two factors, i.e. the availability of the monomer to react with reactive species on the polymer surface, and permeation of the monomer in the bulk of the PP polymer. Unlike AAc, the grafting rate for DMA was proportional to the percentage of monomers present in the reaction medium. Whatever the percentage considered, the degree of DMA grafting was higher than that of AAc.

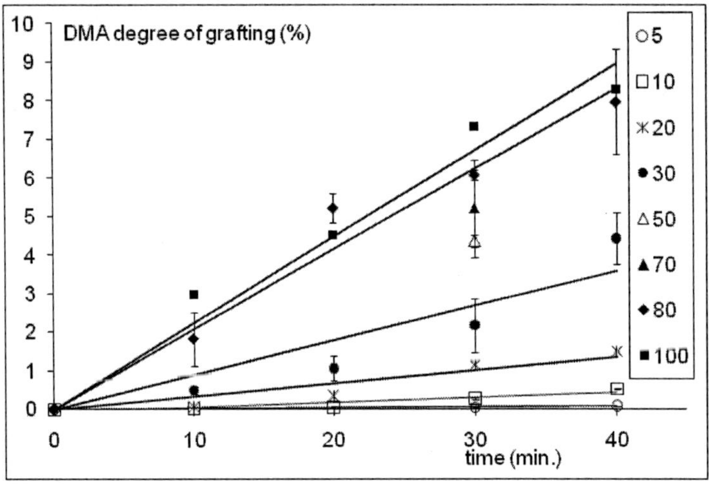

(a)

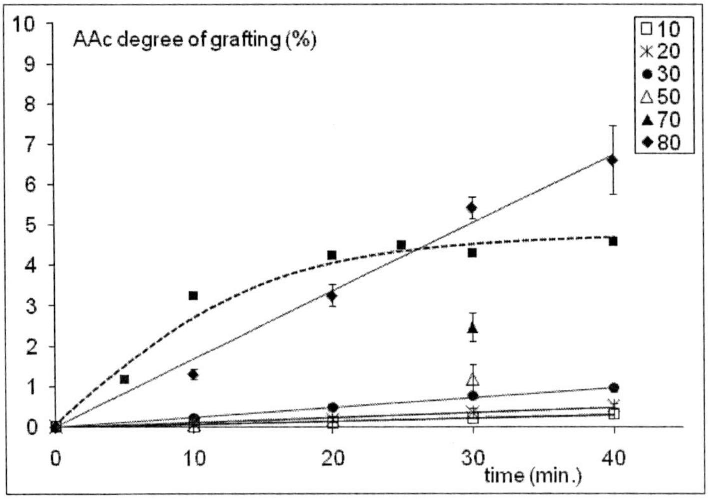

(b)

Figure 1. Variation of the degree of grafting with the reaction time at different monomer concentrations (%). Preirradiation dose 100kGy; temperature 60°C.

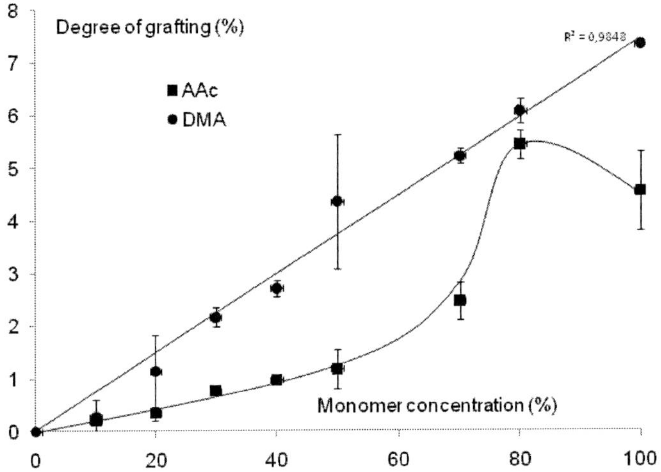

Figure 2. Variation of the degree of grafting with the monomer concentration (%). Preirradiation dose 100kGy; temperature 60°C; reaction time 30 minutes.

Figure 3 presents a comparison of the degree of grafting achieved using either monomers diluted in distilled water (QAS, AAc, DMA), or mixtures in distilled water.

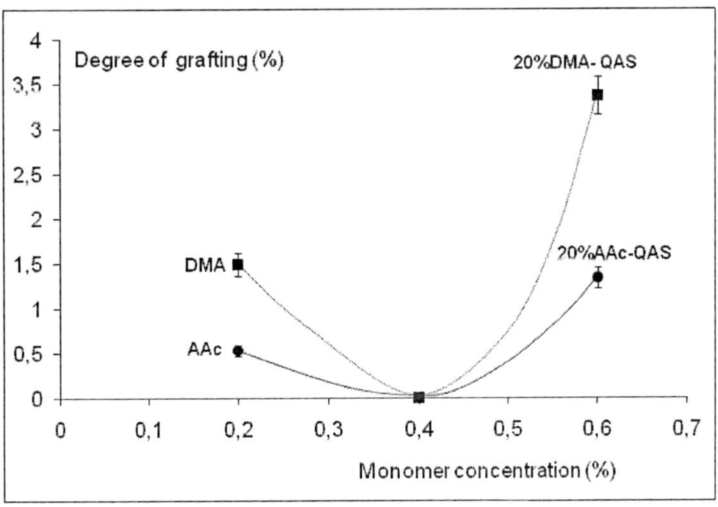

Figure 3. Variation of the degree of grafting with monomers and mixture diluted in distilled water. Preirradiation dose 100kGy; temperature 60°C; reaction time 30 minutes.

The degree of grafting with QAS diluted at 40% was at the limit of the threshold of detection by weighing (10^{-4}g). This was probably due to the viscosity of this compound, its steric hindrance and the surface properties of PP, an apolar and hydrophobic material.

The degrees of grafting with AAc and DMA diluted at 20% were 0.5% and 1.5%, respectively. In the presence of a monomer mix of D2Q4 or A2Q4, it was interesting to observe that the degree of grafting was significantly higher than that obtained using constitutive monomers in the mixture, regardless of the composition of the mix. It thus appeared that the presence of relatively small, highly reactive, hydrophilic compounds might favour the grafting of QAS.

To justify the introduction of functional groups onto the PP, infrared spectroscopy analysis was carried out. The spectra of original and grafted samples were measured directly, without any treatment.

Figure 4a shows the typical FTIR spectra of virgin PP. Absorbance in the 2850–2950 cm^{-1} region corresponds to CH_3, CH_2, CH stretching, while peaks in the 1450–1377 cm^{-1} region are due to CH_2 and CH_3 bending. Various weak, medium and strong peaks in the 809 to 1377 cm^{-1} region correspond to CH_3, CH_2, CH bending, wagging, twisting and C–C stretching. The peaks observed near 1168 cm^{-1} and 999–977 cm^{-1} confirmed the isotactic category of polypropylene used during this study [25, 26].

Figures 4 b, c, d show the essential differences in the spectra of starting and modified PP.

The FTIR spectra of PP-AAc (Figure 4 b) displayed a peak at 1712 cm^{-1} due to carbonyl stretching absorption of the carboxyl group, and the absorption band of OH at 3166 cm^{-1} The characteristic peak of the amide group (1627 cm^{-1}) appeared in the FTIR spectra of PP-DMA samples (Figure 4 c) and the spectra of PP-QAS (Figure 4 d) showed the characteristic absorption bands of the nitrogen group (3300 cm^{-1}), together with the adsorption band of the ester group (1724 cm^{-1}).The presence of these peaks validated the grafting of AAc, DMA and QAS monomers onto the PP.

The FTIR spectra of mixtures displayed the characteristic bands of PP-DMA (or PP-AAc) and PP-QAS, thus confirming formation of the grafting of the binary system.

The general XPS spectra of virgin, exposed and grafted PP provided access to the elemental composition of the molecules on the PP surface, to a depth of approximately 5 nm in the polymer material (Table 1). The surface of virgin PP was composed of 95.3% carbon (C1s at 285eV) and 4.7% oxygen (O1s at 532eV).

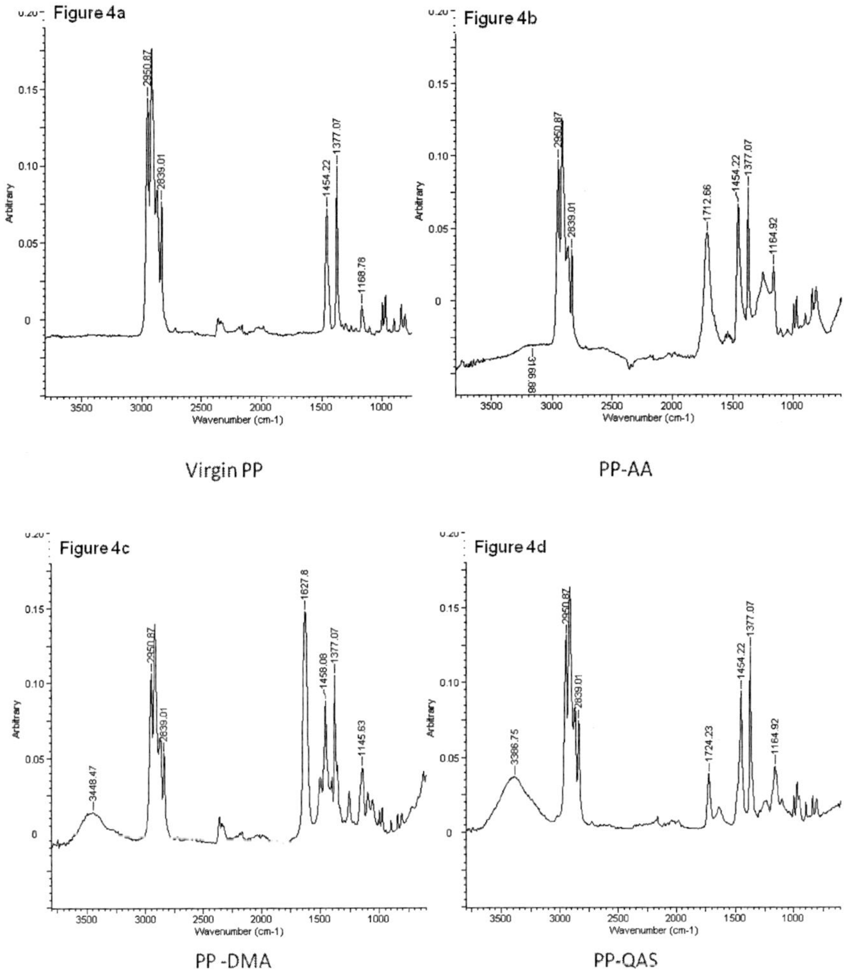

Figure 4. FTIR spectra of ungrafted (a) and grafted PP with AAc (b), DMA (b), QAS (c).

The presence of oxygen could be attributed to thermo-oxidative degradation as well as, partially, to surface oxidation of the PP during storage. By comparison with virgin samples, PP grafted with DMA exhibited a significant reduction in the carbon percentage (95.3→73.7%), a marked increase in the oxygen percentage (4.8→12.5%) and the appearance of nitrogen (N1s peak at 400eV). Because DMA is made up of 71.4% carbon, 14.3% oxygen and 14.3% nitrogen, these variations were consistent with the presence of DMA on the PP surface. And because XPS analyzed the elemental compounds on the surface up to a depth of 5 nm, these results were in favour of DMA mainly being grafted on the surface.

Table 1. XPS elemental analysis of virgin PP, PP grafted with monomers (AAc, DMA, QAS) and different ratios of comonomer mixtures (reaction time: 30 minutes). ***In italics: calculated values***

Sample	$\%C_{1s}$	$\%O_{1s}$	$\%N_{1s}$	$\%Cl_{2p}$	O_{1s}/C_{1s}	N_{1s}/C_{1s}
Virgin PP	95.3 *100*	4.8 *0*			0.04 *0*	
PP grafted AAc	71.4 *60*	27.7 *40*	- -	- -	0.38 *0.66*	- -
PP grafted DMA	73.7 *71.4*	12.5 *14.3*	12.7 *14.3*	- -	0.17 *0.20*	0.17 *0.20*
PP grafted QAS	71.9 *69.2*	17.4 *15.4*	5.8 *7.7*	3.7 *7.7*	0.24 *0.22*	0.08 *0.11*
PP grafted A2Q4	69.8 *65.5*	20.5 *20.6*	5.1 *6.8*	3.1 *6.8*	0.29 *0.31*	0.07 *0.10*
PP grafted D2Q4	72.3 *67.7*	13.6 *16.1*	9.5 *9.7*	4.6 *6.4*	0.19 *0.23*	0.13 *0.14*

After immobilization of the D2Q4 mixture, the appearance of an additional Cl2p peak (200-210eV) was indicative of the presence of QAS. Chlorine corresponds to the counter ion of a quaternary ammonium salt. The slight increase in the O/C ratio and the reduction in the N/C ratio confirmed the fact that the two monomers, DMA and QAS, were grafted onto the surface of PP.

By comparison with non-modified PP, that grafted with AAc displayed a significant reduction in the carbon percentage (95.2→71.4%) and a marked increase in the oxygen percentage (4.8→27.7%). These changes were very probably linked to the presence of AAc on the surface of PP. The O/C ratio of pure AA (0.66) diminished after grafting on PP (0.38). This result suggested that under the grafting conditions applied, and even if most AAc chains were confined to the surface (thus generating a marked increase in hydrophilicity), grafting continued to some extent in the polymer matrix. This AAc depletion at the surface resulted in an increase in the percentage of C1s and a reduction in the percentage of O1s, when compared with the calculated values.

After immobilization of the A2Q4 mixture, the appearance of two additional peaks (N1s and Cl2p) was indicative of the presence of QAS. It is interesting to note that with the A2Q4 or D2Q4 mixtures, the calculated atomic composition of the surface was similar to the experimental atomic composition of the surface. This result may mean that a good correlation existed between the percentage of monomers in the reaction medium and the percentage of grafted monomers.

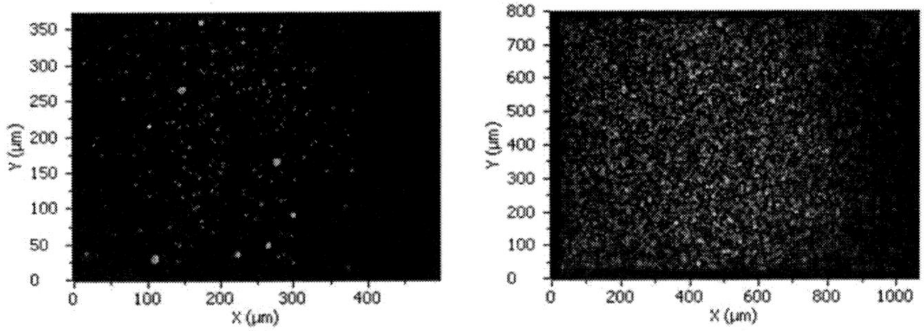

Figure 5. XPS mapping of nitrogen.

Conventional XPS provides an average composition over the analyzed area. However, many sample surfaces are not homogeneous and an element present at apparently low concentrations in the spectrum may be concentrated in one region of the area of interest, possibly leading to a completely erroneous interpretation of the data. We therefore carried out XPS imaging on grafted PP samples (QAS and A2Q4) to map the elemental distribution of nitrogen, as shown in Figure 5. Each resulting image pixel represented the peak height of the imaged nitrogen (N) species at that spatial position. A set of images could thus be quantified pixel by pixel in an analogous manner throughout the spectra, meaning that on the PP sample grafted with the A2Q4 mixture we obtained a surface with an almost homogenous finish of nitrogen compared to PP grafted with QAS. This result indicates that using our experimental system, the grafting of comonomer mixtures occurred in a relatively satisfactory and homogenous manner.

To analyze surface wettability, contact angle measurements were performed on unmodified (virgin and exposed PP) and previously modified (grafted) PP surfaces and the results are presented in Table 2. The water contact angle is the most convenient parameter to analyze the surface hydrophilic/hydrophobic properties of modified polymeric surfaces [27] . Moreover, based on the contact angle it is also possible to determine the energy characteristics of unmodified and modified surfaces using polar and apolar liquids. Three pure liquids, two polar and one apolar, were used to measure contact angles, as described in the experimental part. Most of the liquid used to measure contact angles was capable of interacting via hydrogen bonds, while modified surfaces may interact via polar, apolar and hydrogen bond attractive forces. Interestingly, through these attractive forces, it was possible to verify the interaction behaviour of polar and apolar liquids simultaneously.

Table 2. Contact angles with water for unmodified PP, exposed PP, and PP grafted with AAc, DMA or comonomer mixtures

Monomer	Monomer concentration (%)	Degree of grafting (%)	Contact angle (degree)
Virgin PP	-	-	94 ± 3
Exposed PP	-	-	83 ± 4
PP-AAc	20	0,5	57 ± 7
	100	4,5	40 ±5
PP-DMA	10	0,1	41 ± 5
	20	1.5	34 ± 6
	40	4,3	30 ± 4
PP-SAQ	40	~ 0	25 ± 4
PP-AA/SAQ	20 / 40	1.33	25 ± 11
PP-DMA/SAQ	20 / 40	3.37	24 ± 4

The contact angles of virgin and exposed PP were 94±3° and 83±4°, respectively. However, all the grafted PP displayed a decreasing trend, with an increase in the degree of grafting whatever the ratio of monomers in the reaction medium.

Grafting with pure AAc resulted in an increase in the hydrophilic character of PP when compared with samples prepared using the diluted monomer (20% AAc). These results indicate that graft management is truly a function of the monomer concentration. In the presence of pure AAc, grafted chains of polyacrylic acid were still mainly confined to the surface of the material. This external enrichment tended to increase the surface energy, and hence the hydrophilic character. The effect of external enrichment by polyacrylic acid chains was such that the contact angle decreased to around 40° for a graft level of 4.5%. In the presence of dilute AAc, it is probable that water acted as a carrier agent for monomer permeation within the surface layers. For this reason, most of the grafted substance tended to move inside the layer and little remained on the surface. PP always remained moderately hydrophilic ($\theta \geq 60°$). Moreover, the samples that were grafted with pure monomer were opaque and white, while those grafted using aqueous monomer solutions, such as 20% Acc, were transparent and almost like virgin PP. These results are extremely attractive and give rise to an unique concept in the grafting process where one single consideration of monomer concentration may lead to the surface design and construction of the PP matrix without any significant alterations to the bulk structure. Our findings were supported by the results of other research groups who had found that the grafting of acrylic acid (~2%) led to a minor decrease in contact angle values of up to 80° [28]. These other authors attributed this behaviour to the heterogeneous distribution of grafting at the film surface. However, closer scrutiny of their grafting conditions revealed that the reaction was performed at a monomer concentration of 50%, and most of the grafted material had indeed migrated into the surface layers.

In line with the results obtained by Chen et al. [29], grafting with DMA resulted in a marked increase in the hydrophilic character, whatever the DMA percentage in the reaction medium ($30 \leq \theta \leq 41°$).Probably because of its hydrophilic nature and its molecular volume being higher than AAc (respectively 105.2 and 58.4 $cm^3.mole^{-1}$), the penetration of DMA within the polymer bulk was restricted. Grafting only occurred on the surface of PP, and as a result the contact angle decreased.

After the grafting of QAS, we obtained highly hydrophilic surfaces, but the grafting technique was quite difficult and very little reproducible.

In the presence of a monomer mix of A2Q4 or D2Q4, it was interesting to observe that the degree of grafting was significantly higher than that obtained using constitutive monomers in the mixture, regardless of the composition of the mix. It thus appears that the presence of relatively small, highly reactive, hydrophilic compounds might favour the grafting of QAS [30]. The results concerning water contact angles supported this hypothesis. Indeed, the water contact angle for PP grafted with QAS diluted at 40% in distilled water was ~25°. For samples grafted with AAc or DMA diluted at 20% in distilled water, the contact angles measured were 57° and 34° respectively, while the values for PP samples grafted with A2Q4 or D2Q4 mixes were both 24° (Figure 6). As discussed before, It may be assumed that in the grafted sample QAS was present at the interface of the sample, and AAc (or DMA) was in the sub layers of the surface, which could account for the lower contact angle value following grafting with the comonomer mixture.

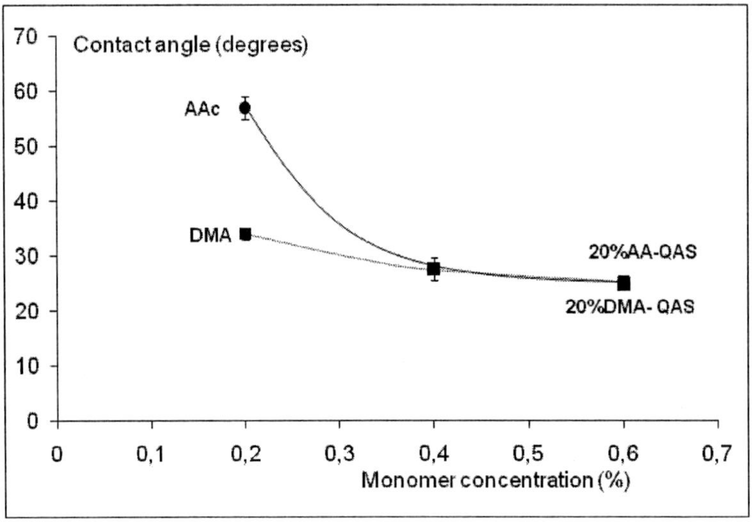

Figure 6. Variation of the contact angle with monomers and mixture diluted in distilled water. Preirradiation dose 100kGy; temperature 60°C; reaction time 30 minutes.

The grafting of hydrophilic monomers on the PP surface resulted in an increase in the Lifshitz-van der Waals and polar components (Table 3).

Table 3. Characteristic surface energy components of ungrafted and grafted PP

Sample	Monomer concentration (%)	γ_{LW} (mJ/m²)	γ_{AB} (mJ/m²)
Virgin PP	-	29.0	0.6
Exposed PP	-	28.2	0.2
PP-AAc	20	37.5	13.2
	100	36. 6	17.8
PP-DMA	10	45.6	4.3
	20	47.7	3.6
	40	47.8	3.4
PP-SAQ	40	33.4	21.1
PP-AAc/SAQ	20 / 40	37.5	16.4
PP-DMA/SAQ	20 / 40	42.9	8.6

For the AAc grafted sample, the γ_{LW} values were close to those of unmodified PP samples. This indicates that there was practically no difference in influence, in terms of van der Waals attractive forces, between AAc grafted and ungrafted samples. Guillemot et al. also reported similar γ_{LW} values (26.3 mJ/m²) for PP [31]. The polar component ranged markedly from 0.6 mJ/m² for unmodified to 17.8 mJ/m²for PP grafted with pure AAc. This suggests that unmodified PP samples were hydrophobic and PP-AAc samples were hydrophilic in nature, although PP samples grafted with AA-20 displayed a moderately hydrophilic character.

Unlike PP-AAc, DMA-modified surfaces displayed a marked capacity for the exchange of Lifshitz-van der Waals interactions, $\gamma_{LW} > 47$ mJ.m^{-2} and a slight increase in the polar component γ_{AB} to >4 mJ.m^{-2}, whatever the monomer concentration.

The grafting of QAS resulted in a very slight modification to the Lifshitz-van der Waals component ($\gamma_{LW} = 33$ mJ.m^{-2}), and a strong increase in the polar component ($\gamma_{AB} = 21$ mJ.m^{-2}).

In all cases, hydrophilicity could be due to the presence of QAS and carboxylic groups on the surface of PP.

Zeta potential is the average potential in the surface and is denoted as ζ. Zeta potential measurements reveal the acidity or basicity of solid surfaces from a quantitative point of view. To investigate the surface charged properties of virgin and grafted PP, streaming potentials were measured as a function of pH in a 1.5×10^{-3} M NaCl solution, followed by calculation of the zeta potential using equation 4.

Figure 7 illustrates the effect of pH on zeta potential values and on the isoelectric points of virgin PP and PP samples with similar surface hydrophilicity $25 \leq \theta \leq 40$. The isoelectric point (iep) is the point at which the zeta potential value is zero, which means that the total positive charges are equal to the total negative charges. Above the iep, the zeta potential increases because of the adsorption of potentially determinant ions, which may also reveal the acid-base nature of the surface from a qualitative point of view.

The iep of virgin PP was 3.5 and the zeta potential value (ζ) was -44mV at the pH of the NaCl 10^{-3} M solution. Although unmodified PP contained no dissociable groups, its ioelectric point was within the acidic range. This could be explained by a preferential adsorption of anions from the NaCl 10^{-3} M solution (OH^- and Cl^-), creating a negative surface charge [32, 33]. Moreover, combined with the results obtained by XPS (4.7% of oxygen on the surface of virgin PP) the negative ζ value may also have been due to O-containing groups such as -OH, -COR or –COOH, that are both polar and acidic. Aranberri-Askargorta et al. had also reported the same iep value (3.3) for pure PP fibres [34].

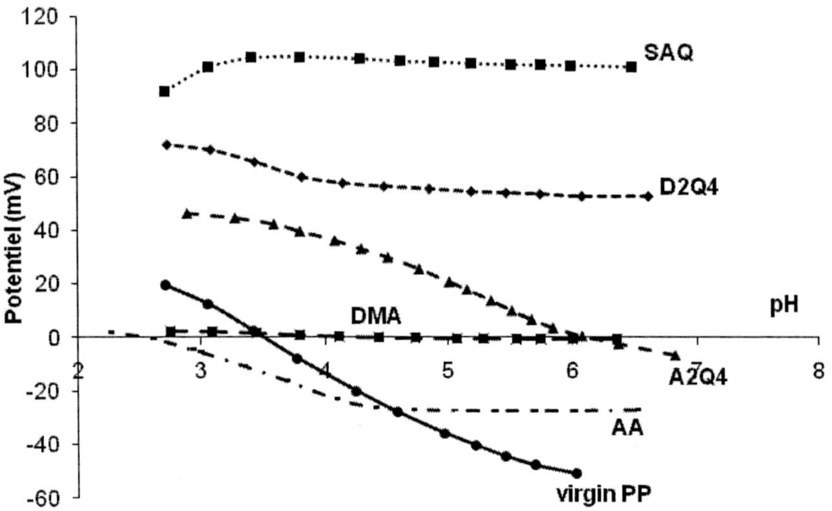

Figure 7. Zeta potential of unmodified and modified PP at different pH values.

The grafting of PP samples caused a marked change to zeta potential values. The iep of PP grafted with pure AAc was shifted to more acidic conditions (2.5) while the zeta potential still remained negative (-27 mV). In that case, the negative charge was mainly due to deprotonation of the carboxylic function (COOH) into carboxylates (COO^-) at a pH > iep. Kuehn et al. had suggested that for polymeric materials, an increase in hydrophilic functional groups on the surface decreases negative zeta potentials as ion adsorption occurs at the same time as water adsorption at hydrophilic sites [35]. However an increase in negative zeta potential may occur if no water adsorption takes place, which causes an increased number of dissociable surface groups[34]. By contrast, the iep of PP grafted with QAS and with comonomer mixtures (A2Q4, D2Q4) was shifted to more basic conditions and the surface charges were positive at physiological pH. Nitrogen could be considered as the source of the positively charged surface. The zeta potential *vs.* pH in Figure 7 clearly showed that the surface charge was positive throughout the entire pH range (2-7) when PP was grafted with QAS and D2Q4, so that these surfaces did not display any iep. Presumably, an excess of basic groups was responsible for the positive charge revealed at the surface of PP modified with A2Q4 and D2Q4. They may have had a higher density of N-containing groups ($-N^+R_3$) than O-containing groups (-COR or -COOH). The magnitude of ζ was reduced from -44 to 0.3 mV after grafting with DMA. This neutral value was probably linked to the -COR groups present on the surface. These groups being less polar and less acidic than –COOH, the grafting of DMA resulted in a less negative charge than PP grafted with AAc.

It is worth pointing out that with a similar surface hydrophilicity ($25 \leq \theta \leq 40$), the zeta potential and isoelectric point varied as a function of the monomer or comonomer mixtures. It is thus highly likely that the nature of the constitutive monomers in the mixes, and their distribution, conditioned the surface properties of the material (Figure 8).

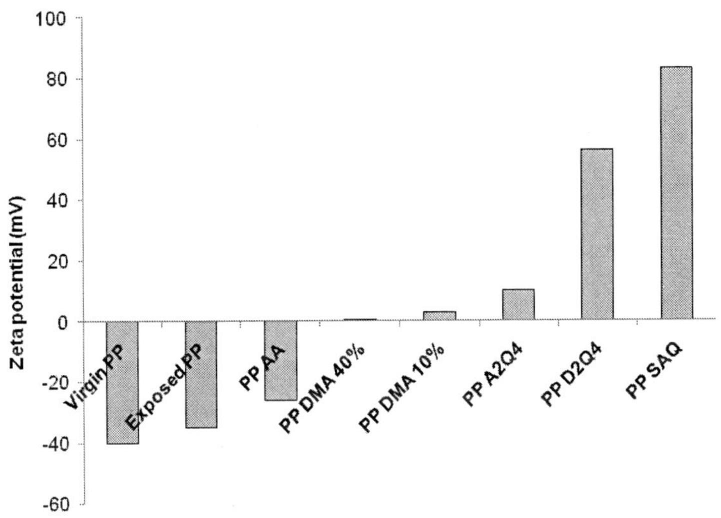

Figure 8. Variation of the Zeta potential as a function of the monomer or comonomer mixtures.

The results shown in Table 4 and Figure 9 underline the influence of the rate of grafting on the surface topography. At extremely low rates of grafting (with the limit of the threshold of detection by weighing) the roughness of micrometric PP (PP_M) or nanometric (PP_N) was not modified (0.25 and 0.04 µm, respectively). On the other hand, as from 0.1% of grafting, the roughness of PP_N increased significantly (0.04 to 0.08 µm) while that of the PP_M (0.25 µm) remained unchanged for as long as the rate of grafting remained lower than 0.2%. For grafting rates higher than 0.2% there was a linear relationship between the degree of grafting and the roughness of the surface (Figure 10).

The increase in roughness related to the grafting was more significant on nanometric substrates than on micrometric substrates, whatever the rate of grafting.

Table 4. Variations in surface topography according to the rate of grafting

Micrometric PP		Nanometric PP	
Degree of grafting (%)	Roughness (µm)	Degree of grafting (%)	Roughness (µm)
~ 0	0.25	~ 0	0.04
0.02	0.25	0.01	0.05
0.21	0.26	0.04	0.04
0.82	0.49	0.09	0.08
1.27	0.65	3.24	3.02
4.32	1.74		

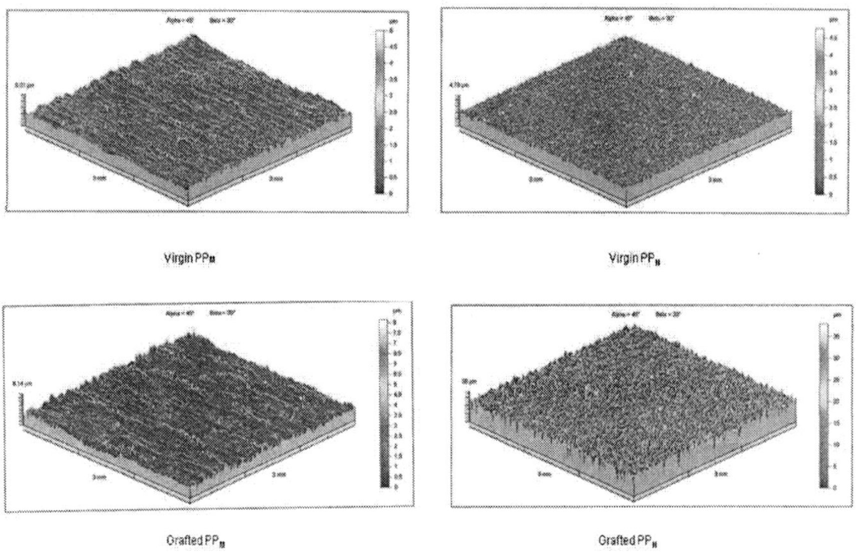

Figure 9. Surface topographies of virgin micrometric or nanometric PP (PP_M , PP_N) and PP_M or PP_N grafted with respectively 0.8% and 3.2 % of DMA.

A scanning electron microscope (SEM) was used to visualize the morphologies of PP samples before and after grafting (Figure 11). It could be seen that the surface of unmodified PP_M displayed some parallel strips not observed on PP_N, which looked smooth (Figure 11 a, b). This result, already reported in the literature [36], most probably resulted from the shape of the PP sheets produced by injection. Ionizing treatment did not seem to affect the surface topography of native material whether it was micrometric or nanometric (Figure 11 c, d). On the other hand, it was interesting to note that the grafting of hydrophylic monomers rigorously followed the surface topography of the substrate concerned and resulted in a very different surface structure, depending on whether a PP_M or a PPN was considered.

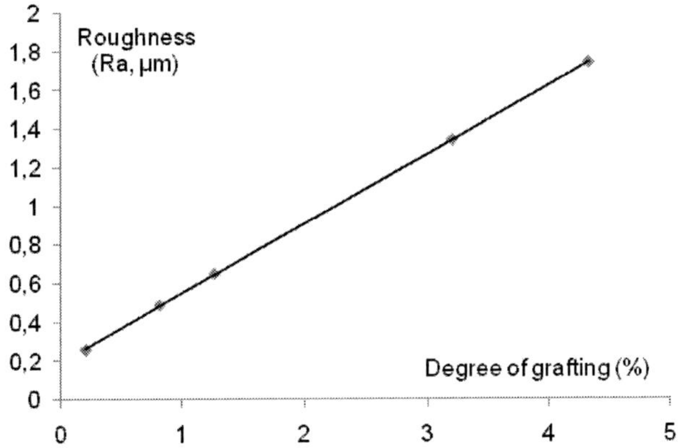

Figure 10. Variation of the roughness of micrometric PP with the degree of grafting.

A marked increase in the relief of the parallel strips was observed after grafting on a micrometric substrate. This phenomenon was especially significant when the rate of grafting was high (Figure 11e, g). In the case of PP_N, at grafting rates of about 0.1% the surface of the material displayed a relief with an "orange peel" structure (Figure 11 f). A high rate of grafting (3%) tended to affect this structure (Figure 11h).

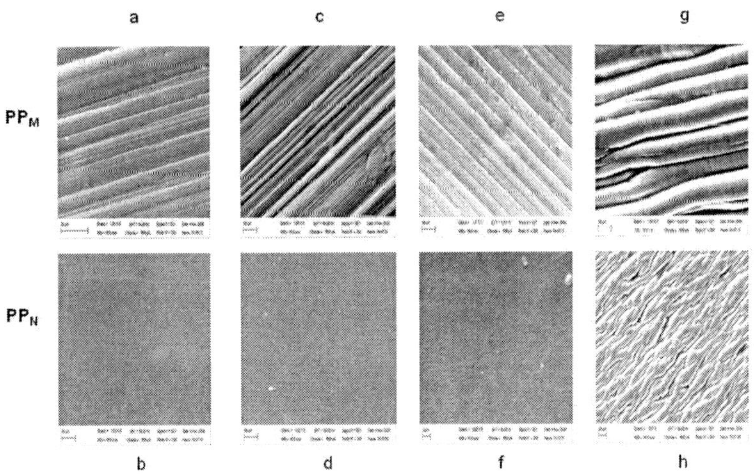

Figure 11. Scanning Electron microscopy images of micrometric PP_M (on the top) and nanometric PP_N (at the bottom); a, b unmodified PP; c,d exposed PP; e,f grated PP with diluted DMA (10%) ; g,h grafted PP with diluted DMA (40%).

Prior to bacterial adhesion, tests of particle adhesion (2 μm diameter polystyrene or silica beads) were performed on native and modified PP surfaces in order to clarify the mechanisms involved in adhesion phenomena. Indeed, these beads constitute a less complex model to understand adhesion phenomena insofar as their surface properties do not vary during contact with a surface, unlike micro-organisms.

The contact angle method, coupled with the Young-van Oss equation (equation 2) made it possible to determine the energy characteristics of the silica and PS beads used during adhesion tests (Table 5). It can be noted that, as expected, the commercial PS beads were more hydrophobic than the silica beads. PS beads also displayed a slightly lower polar character (3.8 instead of 6.3 mJ/m²) and a similar capacity for the exchange of Lifshitz-van der Walls interactions (γ_{LW} ~40mJ/m²). The zeta potential of PS beads at the pH of the NaCl solution at 1.5 x 10-3 was -25 mV.

For the adhesion tests, PP samples were immersed in a solution of beads at a concentration of 5×10^{-3}% diluted in an NaCl solution at 1.5×10^{-3} M for one hour (silica beads) and three hours (PS beads).

Table 5. Energy characteristics of silica and poystyrene beads used in adhesion tests

	Silica beads	Polystyrene beads
θ water (°)	63 ± 8	108 ± 2
θ formamide (°)	51 ± 7	93 ± 5
θ dioodomethane (°)	51 ± 5	54 ± 3
γ_{LW} (mJ/m²)	33.6	31.9
γ_{AB} (mJ/m²)	6.1	3.8
$\gamma_{T'}$ (mJ/m²)	39.7	35.7

With γ_T energy of surface, γ_{AB} and γ_{LW} the polar and Lifshitz-van der Waals components of surface tension respectively; θ is the contact angle in pure liquids of water, formamide and diiodomethane.

As during the grafting stages, optical microscope observations appeared to indicate that the adhesion of PS particles was governed by surface topography. On PP_M, the beads aligned themselves along the ridges, while they did not adopt any preferential orientation on PP_N (Figure 12). In order to confirm or refute these observations, an auto-correlation analysis was performed. This mathematical tool enables a comparison of the relative positioning of particles and the tracing of curves that correspond to their preferred alignments.

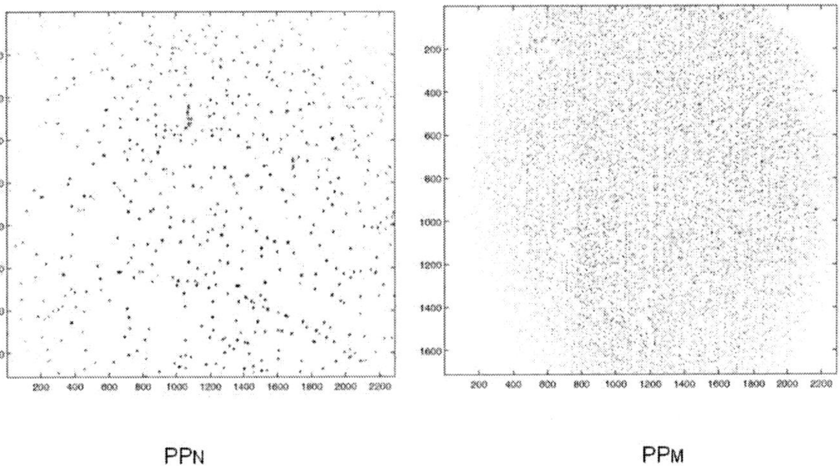

Figure 12. Images of adherent PS beads observed under an optic microscope (BX51, Olympus) though a 40X objective for micrometric PP_M (b) and nanometric PP_N (a).

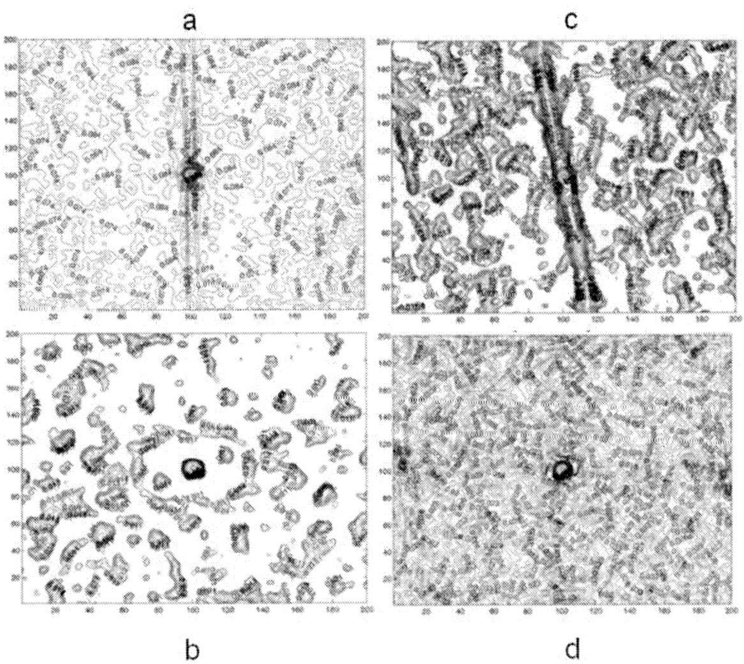

Figure 13. Images of auto correlation analysis of PS beads on unmodified PP_M (a), grafted PP_M with 40% diluted DMA (c), unmodified PP_N (b) and grafted PP_N with 40% diluted DMA (c).

The results presented in Figure 13 confirmed the preferred linear orientation of PS beads on the surface of native micrometric PP (Figure 13 a) or of PP modified by granting with 40% DMA, i.e. a surface with more marked roughness (Ra 1.74 μm) (Figure 13 c). By contrast, on nanometric substrates, bead adhesion did not follow any preferential direction, whatever the degree of roughness (Figure 13 b, d). Similar behaviour was observed with the silica beads (Figure 14). However, under monitored acquisition conditions, the adhesion mode of the beads did not appear to influence the number of beads adhering to the substrate.

If surface roughness was considered, it could be seen that the percentage coverage of PP by PS beads was always greater when the surface was rough. By contrast, regarding their hyodrophilic/hydrophobic character, PS beads displayed much greater affinity for hydrophobic control substrates (θ=95°) than for highly hydrophilic modified substrates (θ=40°, 30° or 20°) whatever the degree of surface roughness (Figure 14).

Figure 14. Images of adherent silica beads observed under an optic microscope (BX51, Olympus) though a 40X objective for micrometric PP_M grafted with 3.2 % of DMA.

Using silica beads, the percentage coverage of hydrophilic substrates was significantly higher (factor ~15) than that seen on control substrates. Crossing of these results suggested that the wettability of the surface had a predominant effect on particle adhesion, while topography (roughness) exerted only a secondary effect.

With respect to short and long distance interactions, PP-modified with DMA displayed characteristics similar to those of polystyrene beads. Thus no attractive or repulsive interactions could be generated. By contrast, electrostatic interactions were attractive between QAS-modified surfaces (ζ = 102 mV) and polystyrene beads (ζ = -25 mV), thus favouring the approach of beads to the surface of PP. In principle, the influence of Lewis acid-base interactions is much greater than that of electrostatic interactions. However, these are short distance interactions, and it is necessary for the bead and the substrate to be relatively close for these forces to become effective. These conditions were probably not achieved between the polystyrene beads and QAS- modified surfaces (Figure 14).

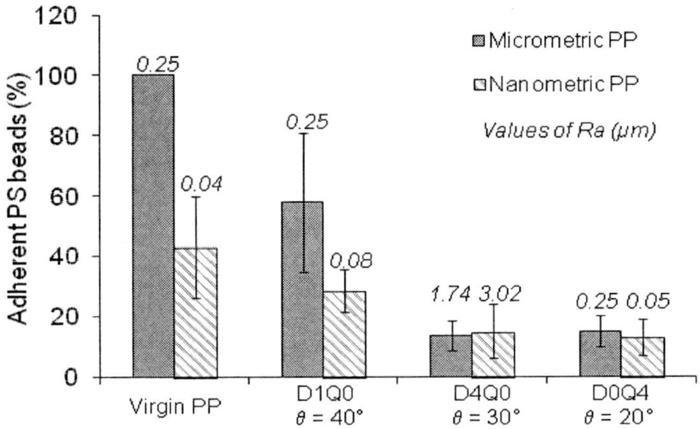

Figure 15. Variation of the percentage adhesion of PS beads with the roughness of unmodified and modified PP surfaces.

The initial adhesion tests performed by sedimentation for 24 hours at 20°C with *Listeria monocytogenes* (LM) demonstrated a difference in the bioadhesive behaviour of the strain employed, as a function of the physicochemical properties of the substrate surface tested.

On QAS-grafted PP substrates, microbiological analyses demonstrated an increase of around 2 log10 CFU/cm² when compared with the control PP, whatever the degree of surface roughness. These results suggest that the hydrophilicity of the surface was not the only parameter influencing LM adhesion. Attractive interactions between QAS-grafted materials (ζ =+ 102 mV) and LM (ζ = - 41 mV) thus favoured bioadhesion.

CONCLUSION

The aim of these studies was to modify the surface of PP at a low grafting level and to introduce hydrophilicity by using a mixture of two monomers, a quaternary ammonium salt (QAS) and acrylic acid (AAc) or N,N-dimethylacrylamide (DMA), so that these surfaces could then be analyzed for bacterial adhesion.

The grafting rates and physicochemical characteristics of polypropylene surfaces functionalised by the radio-grafting of hydrophilic monomers were dependent on the operating conditions, such as the ionisation dose, the duration of exposure or temperature. When these conditions were perfectly controlled, it was possible to develop highly hydrophilic surfaces (water contact angle $25° < \theta_{water} < 35°$) with excellent reproducibility.

The grafting of DMA on a PP substrate resulted in a hydrophilic surface whatever the percentage of monomer in the reaction medium and the degree of grafting. These results constitute an unique concept in the grafting process where just a single aspect of the monomer concentration can design the surface and construct the PP matrix without this involving any significant alteration to the bulk structure. A hydrophilic surface was also obtained after grafting with AAc, but only at high degrees, and it is well known that such a high degree of grafting affects the physical structure of PP, and notably its cristallinity.

The cografting of a highly reactive, hydrophilic monomer such as acrylic acid (AAc) or dimethyl acrylamide (DMA), ensured "prehydrophilization" of the material which in turn favoured the homogeneous grafting of QAS .

The grafting of an AAc or DMA monomer, or a D2Q4 or [AAc2Q4 comonomer, markedly modified the hydrophilicity of the PP surface. For the same surface hydrophilicity, and depending on the nature of the monomer or comonomer grafted, it is thus possible to establish a gradient of electrostatic and polar properties on the surface of plastic materials. In view of the uncertainties regarding the homogeneity of micro-organisms in a microbial population under field conditions, the bioadhesive impact of micro-organisms on surfaces with a controlled gradient of properties will undoubtedly help to define the best alignment between surface properties and antimicrobial efficacy.

Grafting led to significant changes to structural morphology, particularly when the initial substrate was smooth. However, under monitored acquisition conditions, the mode of adhesion did not seem to affect the number of particles that had adhered, whatever their nature (silica or PS beads).

Image analysis of the adherent particles revealed that surface hydrophilicity played a major role in particulate adhesion, while topography (roughness) only had a secondary effect. However, electrostatic interactions appeared to predominate in the bioadhesion mechanism.

More studies will be performed to validate these results, using micro-organisms such as *Escherichia Coli* , *Listeria Monocytogenes* or *Pseudomonas Aeruginosa* which are responsible for many, more or less severe outbreaks of food poisoning.

REFERENCES

[1] A. Terada, A. Yuasa, S. Tsuneda, A. Hirata, A. Katakai, M. tamada, *Colloids Surf., B Biointerf.* 43 (2005) 99.

[2] R. Bos, H.C.V.D. Mei, H.J. Brusscher, *FEMS Microbiol. Rev.* 23 (1999) 179.

[3] P.J. McClure, A.L. Beaumont, J.P. Sutherland, T.A. Roberts, *International Journal of Food Microbiology* 34 (1996) 221-232.

[4] K. Kim, J. F. Frank, *Journal of food protection* 8 (1994) 720-726.

[5] A.A.Mafu, D.Roy, J. Goulet, P. Magny, *Journal of Food Protection*, 53 (1990) 742-746.

[6] A.A. Mafu, D. Roy, J. Goulet, L. Savoie, *Applied and Environmental Microbiology*, 57 (1991) 1969-1973.

[7] M. Hermansson, *Colloids Surf. B Biointerf.* 14 (1999) 105.

[8] C.J. van Oss, R.J. Good, M.K. Chaudhury, *J. Colloid Interf. Sci.* 111 (1986) 378.

[9] C.J. van Oss, *Interfacial Forces ; Aqueous Media*, Marcel Decker, New York: (1994).

[10] P.R. Rutter, B. Vincent, The adhesion of microorganisms to surfaces, physico-chemical aspects; *Microbial Adhesion to Surfaces*, Ellis Horwood, London (1980).

[11] L. Massi, F. Guittard, S. Gériabaldi, R. Levy, Y. Duccini, *Int. J. Antimicrobiol. Age.* 21 (2003) 20.

[12] V. Majtan, L. Majtanova, *Chem. Papers* 54 (2000) 49.

[13] F. Kopecky, *Pharmazie* 51 (1996) 135.

[14] M.V. Jones, T.M. Herd, H.J. Christie, *Microbios* 58 (1989) 49.

[15] R. Briandet, *Thesis, ENSA-RENNES*, France, (1999).

[16] L. Boulangé-Petermann, A. Doren, B. Baroux, M.N. Bellon-Fontaine, *J. Colloid Intrerf. Sci.* 171 (1995) 179.

[17] A. Vernhet, J.Y. Leveau, O. Cerf, M.N. *Bellon-Fontaine, Biofouling* 5, (1992) 323.

[18] R.A.V. Wagenen, J.D. Andrade, *J. Colloid Interf. Sci.* 76 (1980) 305.

[19] Anjum, N.; Moreau, O.; A.M. Riquet; *J Appl Polym Sci,* In Press, 2010.

[20] A. Wirsen, AC Albertsson, *J. Polym. Sci.A*, 33(1995) 2049-2055

[21] O. Ohrlander, A. Wirsen, AC Albertsson, *J Appl Polym Sci*, 37 (1999) 1643-1649

[22] C.J. van Oss, *Colloids Surf. B Biointerf.* 5 (1995) 91.

[23] N., Anjum, O., Moreau, A.M.,Riquet, *J Appl Polym Sci*, 100 (2006) 546-558

[24] N., Anjum, M.-N. Bellon-Fontaine, J.-M. Herry, A.-M. Riquet, *J Appl Polym Sci*, 109 (2008) 3, 1746-1756

[25] D. Sinha, T. Swu, S.P. Tripathy, R. Mishra, K.K. Dwivedi, D. Fink, *Rad. Eff. Deffect Sol.* 158 (2003) 531

[26] N.L. Mathakari, D. Kanjilal, V.N. Bhoraskar, S.D. Dhole, *Nuclear Instruments and Methods in Physics Research* B 266 (2008) 1793–1798

[27] C.-.L. Li, C.-Y. Tu, J.-S. Huang, Y.-L. Liu, K.-R. Lee, J.-Y. Lai, *Surf. Coat. Technol.* 201 (2006) 63.

[28] I.Louis Joseph Dogue, N.Mermilliod, A.Gandini, *J Appl Polym Sci*, 56 (1995) 33.

[29] C.P. Chen, B.T. Ko, S.L. Lin, M.Y. Hsu, C. Ting, *Polymer* 47 (2006) 6630.

[30] F. Liu, B.K. Zhu, X. You-Yi, *Appli. Surf. Sci.* 253 (2006) 2096.

[31] G. Guillemot, G. Vaca-Medina, H. Martin-Yken, A. Vernhet, P. Schmitz, M. Mercier-Bonin, Colloids Surf., B Biointerf. 49 (2006) 126.

[32] P. Weidenhammer, H. J., Jacobasch, *J. Colloid Interface Sci.* 180 (1996) 232-236

[33] H. J., Jacobasch, F. Simon, P. Weidenhammer, *Colloid Polym Sci.* 276 (1998) 434-442

[34] Aranberri-Askargorta, T. Lampke, A. Bismarck, *J. Colloid Interf. Sci.* 263 (2003) 580.

[35] N. Kuehn, H.-J. Jacobasch, K. Lunkenheimer, *Acta Polym.* 37 (1986) 394.

[36] C. Y. Huang, C. L. Chen, *Surface and coatings technology* 153, (2002) 194-202

In: Polystyrene: Properties, Performance... ISBN: 978-1-61209-117-4
Editor: James E. Gray

Chapter 4

DIRECT FLUORINATION OF POLYSTYRENE

*A. P. Kharitonov**

Institute of Energy Problems of Chemical Physics (Branch) of the Russian Academy of Sciences, Chernogolovka, Moscow region, Russia

ABSTRACT

In this paper, fundamental features of the direct fluorination of polystyrene is reviewed. Direct fluorination of polymers (i.e. treatment of a polymer surface with gaseous fluorine and its mixtures) proceeds at room temperature spontaneously and can be considered as a surface modification process. A large variety of experimental methods, such as FTIR spectroscopy, visible and near UV spectroscopy, Electron Spin Resonance spectroscopy, laser interference spectroscopy, refractometry, etc. was applied. Fundamental features of the direct fluorination, such as influence of treatment conditions (composition of the fluorinating mixture, fluorine partial pressure, fluorination duration) on the rate of formation, chemical composition, density, refraction index of the fluorinated layer, formation of radicals during fluorination and their termination were studied. On the base of obtained experimental data a simple theoretical model of the direct fluorination of polymers was developed.

---

* khariton@binep.ac.ru

INTRODUCTION

Commonly used polymers have a lot of advantages, such as a low cost, processability, but they have also a lot of disadvantages (often poor adhesion, poor printability and barrier properties, low chemical resistance, etc.. It is possible to fabricate the whole article from specialty polymers, e.g. fluorine-containing polymers, which have improved commercial properties. Fluorinated polymers have a set of unique properties such as enhanced chemical stability, thermostability, good barrier properties, etc. [1-13]. However practical use of specially synthesised polymers, such as fluorine-containing polymers, is restricted due to their high cost and complexity of synthesis. But very often application properties of polymer goods are defined mainly by their surface properties. Hence it is not necessary to fabricate articles from fluoropolymers but simpler, cheaper and more convenient to apply a surface treatment of articles made from commonly used polymers. In this case the direct fluorination can be effectively used [1-13]. Direct fluorination of polymers is a heterogeneous reaction of gaseous F_2 and its mixtures with a polymer surface. This is a method of the surface modification: for majority of glassy polymers only upper surface layer is modified (<0.01-10 μm in thickness), but the bulk properties remain unchanged. Since fluorination is one of the most effective chemical methods to modify and control physicochemical properties of polymers over a wide range, this process has become an important tool of great interest. Direct fluorination has many advantages when used in industry. Due to the high exothermicity of the main elementary stages fluorination proceeds spontaneously at room temperature with sufficient for industrial applications rate. Direct fluorination is a dry technology. Polymer articles of any shape can be treated. There are safe and reliable methods to neutralize (by converting into the solid phase) unused F_2 and the end-product HF. These features of the direct fluorination initiated its wide industrial utilization: for enhancement of the barrier properties of automotive polymer fuel tanks and vessels for storage of toxic and volatile liquids, improvement of gas separation properties of polymer membranes, enhancement of adhesion properties of polymer articles and reinforcement of polymer composites, improvement of antibacterial and chemical resistance [1-13].

MATERIALS AND METHODS

Fluorine contained less than 0.1% of admixtures (mainly oxygen). He, Ar, N_2 and O_2 were of 99.99-99.999% purity.

A large variety of common experimental methods, such as FTIR, visible and near UV spectroscopy, Electron Spin Resonance spectroscopy, refractometry, etc. was applied. Also an original laser interference methods was especially elaborated to measure the dependence of the thickness of fluorinated layer on fluorination duration "in situ".

EXPERIMENTAL RESULTS AND DISCUSSION

The visible spectra of PS films treated not through all its thickness exhibit interference features similarly to the visible spectra of a lot of glassy polymers [5-11] and consist of a set of equidistance (in wavenumber scale) maxima and minima (Figure 1). It is due to the following reason: fluorine treated free polymer film treated from both sides simultaneously represents a three-layer structure and consists of substantially (in many cases- practically totally fluorinated) fluorinated layers and virgin (unmodified) layer, which are separated by a very narrow transient reaction zone (Figure 2). The thickness δ_b of a transient zone is much less than 0.1 μm [9, 11, 14-15]. The main chemical conversion processes proceed inside that transient reaction zone. But the main chemical composition changes may be followed by several post-reactions inside the fluorinated layer. The rate of formation of fluorinated layer is limited by the rate of penetration of fluorine through fluorinated layer to untreated one. Hence the rate of formation of fluorinated layer is limited by the permeability $\overline{P}_{F2}$ of fluorine through fluorinated polymer.

The existence of a sharp boundary between virgin and fluorinated layers made possible application of an original method (called laser interference method) to measure the dependence of the thickness of fluorinated layer on time "*in situ*" on the surface of both optically transparent and deposited onto nontransparent supports polymer films [6-15]. Polymer film with very flat ("shining") surface was placed inside the reaction vessel, equipped with 2 sapphire optical windows. Intensity of monochromatic light (He-Ne laser was used as a light source, λ= 0.6328 μm), reflected from the polymer film surface, was monitored in the course of fluorination. If the treated polymer film consists of a fluorinated layer and unmodified one, which are separated by a very narrow transient boundary reaction zone, and the thickness $\boldsymbol{\delta_b}$ of that boundary zone is $\boldsymbol{\delta_b} \ll \boldsymbol{\lambda}/(4 \cdot \boldsymbol{n_F}) \ll 0.1$ μm *($\boldsymbol{n_F}$* is the refraction index of fluorinated polymer and usually $\boldsymbol{n_F}$ is close to 1.36-1.45 [9, 11] and is smaller than the refraction index of virgin polymer), then monitored intensity represents a set of consequent maxima and minima (Figure 3). The first minimum corresponds to the thickness of fluorinated layer $\boldsymbol{\delta_F} = \boldsymbol{f} \cdot \boldsymbol{\lambda}/(4 \cdot \boldsymbol{n_F})$ (so-called "antireflecting layer"), the first maximum- to $\boldsymbol{\delta_F} = \boldsymbol{f} \cdot 2 \cdot \boldsymbol{\lambda}/(4 \cdot \boldsymbol{n_F})$, the second minimum- to $\boldsymbol{\delta_F} = \boldsymbol{f} \cdot 3 \cdot \boldsymbol{\lambda}/(4 \cdot \boldsymbol{n_F})$ and so on. Such a structure of a reflected light is due to the interference between two light beams, reflected from the surface fluorinated polymer- gas phase and reflected from the surface fluorinated polymer- unmodified polymer. Coefficient $\boldsymbol{f} = [1 - (\sin \boldsymbol{\beta})^2 (\boldsymbol{n_F})^{-2}]^{-0.5}$ depends on the angle $\boldsymbol{\beta}$ of the incidence of the probe light beam. Usually $\boldsymbol{\beta}$ was equal to 45° and respectively $\boldsymbol{f}$=1.16 Such a method can be applied from $\boldsymbol{\delta_F}$~0.1-0.12 μm. In some experiments $\boldsymbol{\delta_F}$ was monitored up to 50 μm.

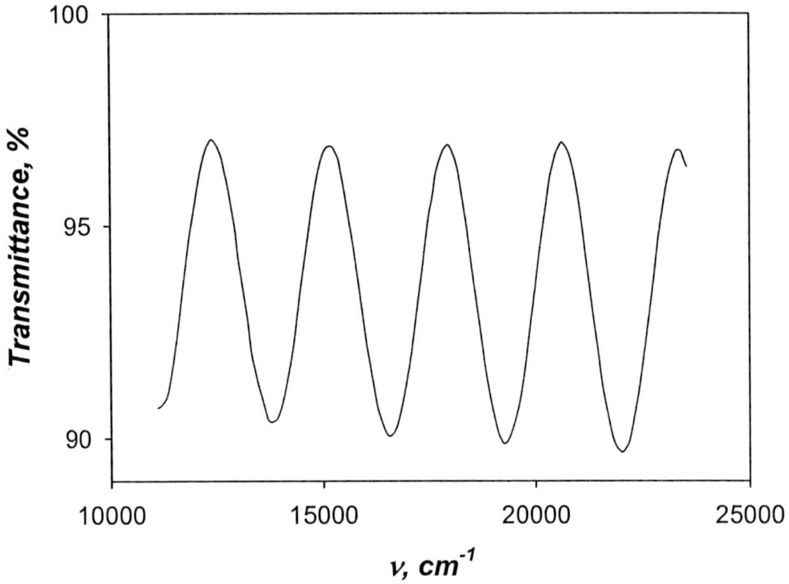

Figure 1. Visible spectrum of the PS film fluorinated from both sides. Thickness of fluorinated layer from each side $\boldsymbol{\delta_F}$=1.3 μm.

Another method can be applied to measure the $\boldsymbol{\delta_F}$ value. The transition (or absorption) spectra of fluorinated polymer films in the visible and near UV spectral regions have interference features and consist of a set of equidistant minima and maxima when frequency ν is measured in cm^{-1} scale (see Figure 1) and the δ_F value can be calculated by the following formula: $\delta_F=(2.n_F.\ \Delta\ \nu)^{-1}$, where $\Delta\nu$ (cm^{-1}) is the distance between two neighboring interference spectrum maxims (minima).

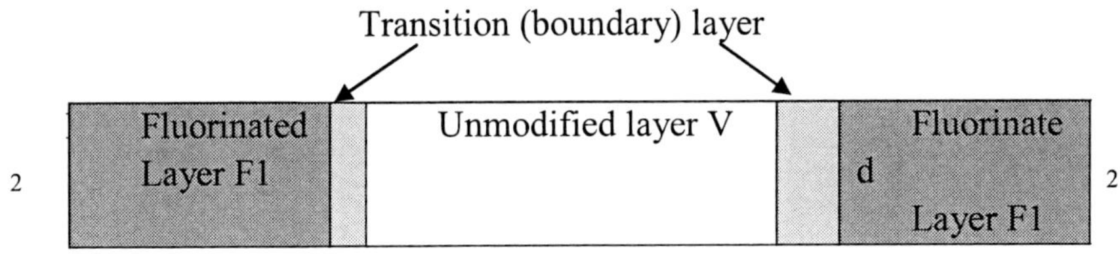

Figure 2. Cut of a partially fluorinated polymer film, treated with fluorine from both sides.

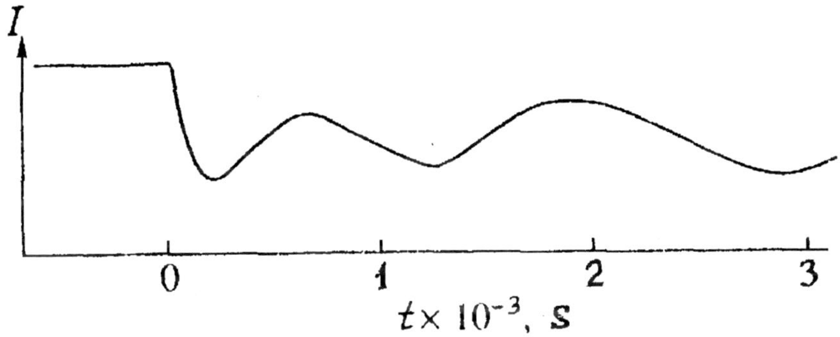

Figure 3. Typical dependence of the intensity ***I*** of monochromatic light (λ= 0.6328 μm), reflected from the polymer film surface, on duration of fluorination ***t*** monitored in the course of fluorination. Fluorination was started at t=0.

For thinner layers such a situation was not observed because the "*in situ*" laser interference method of the measurement of the kinetics of formation of thickness of fluorinated layer (see below) can be applied for the case when $\delta_F \geqslant 0.1$ μm. That "*in situ*" laser interference method allowed the estimation of the dependence of the upper limit of the thickness δ_b of transition layer between fluorinated and untreated polymer layers [14-16]:

Refractive index value is needed to calculate the thickness of fluorinated layer by two abovementioned methods. The dependence of the refractive index n_F (measured at sodium D-line λ=0.593 μm) of PS treated with F_2+O_2 mixture vs $p_0/(p_0+p_F)$, where p_0 and p_F are oxygen and fluorine partial pressures in fluorinating mixture, is presented in the Figure 4. It is evident that n_F value is slightly increased with oxygen concentration in fluorinating mixture from 1.366 ($p_0/(p_0+p_F)$=0) to 1.4570 ($p_0/(p_0+p_F)$=0.75). Refractive index of virgin PS was equal to 1.59.

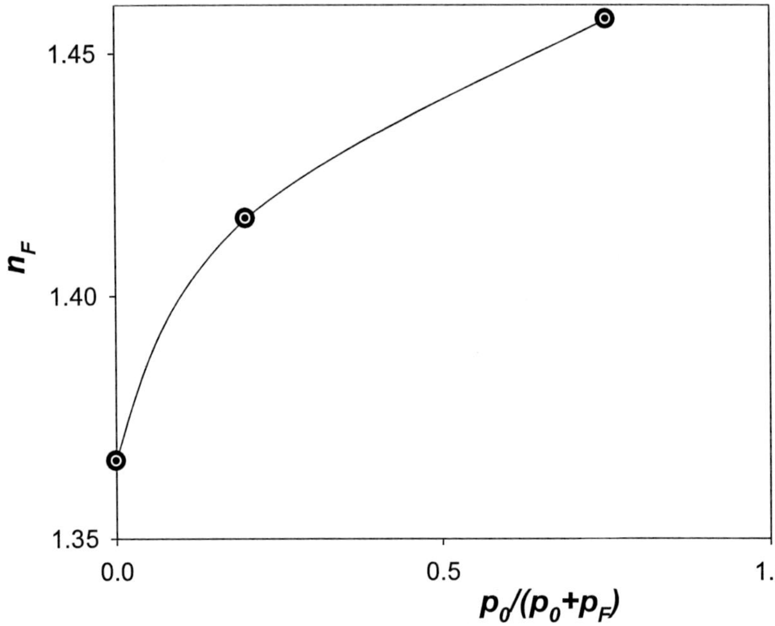

Figure 4. Refractive index n_F (measured at sodium D-line λ=0.593 μm) of PS treated with F_2+O_2 mixture vs $p_0/(p_0+p_F)$, where p_0 and p_F are oxygen and fluorine partial pressures in fluorinating mixture. Fluorine partial pressure was maintained constant, p_F=0.2 bar, treatment temperature was equal to 294 K.

Investigations of the kinetics of growth of the thickness δ_F of PS fluorinated layer was previously performed in [17]. δ_F (measured on the base of interference features in the visible spectra of fluorinated films) for the case of PS was proportional to the square root of the fluorination duration. The main disadvantage of that method is as follows: fluorination must be interrupted to carry out each measurement of δ_F. Estimate of the thickness δ_b of transient layer between fluorinated and untreated layers made in [17] was unbelievable great - more than several tens microns. Unreliability of those results becomes evident taking into account that the thicknesses of fluorinated layer did not exceeded several microns and applied method could be used only if the thickness of the boundary layer did not exceed ~λ/4~0.1-0.15 μm. According to estimations made in [18] the thickness δ_b of the transient reaction zone was as follows: $\delta_b < 0.004 \cdot \delta_F$ when $\delta_F > 1$ μm and $\delta_b << 0.1 \cdot$μm when $\delta_F > 1$ μm.

In all the experiments a pellet of NaF was inserted into the reaction vessel to remove HF via the following reaction: $NaF + HF \rightarrow NaHF_2$ (solid) because HF should inhibit the fluorination rate. The described above laser interference method was used. For all used fluorinating mixtures δ_F value was proportional to the square root of fluorination duration t (Figure 5):

$$\delta_F = A \cdot t^{0.5} + \text{const} = B \cdot (p_F)^k \cdot t^{0.5} + \text{const} \qquad (1)$$

where A value in principe depends on p_F, p_0, p_{He}, p_N, и p_{HF} – partial pressures of F_2, O_2, He, N_2 and HF. «const» value in (1) can be neglected for all the studied fluorinating mixtures. It was shown that the rate of formation of fluorinated layer depends on the fluorine partial pressure and does not depend (within 5-10% accuracy) on the presense of N_2, He, Ar and CO_2 when their concentration in fluorinating mixture varies from 0 up to 88 volume %. For the treatment with mixtures of fluorine with N_2, He, Ar and CO_2 at temperature 293±1 K

$$\delta_F = 0.14 \cdot (p_F)^{0.68} \cdot t^{0.5} \qquad (2)$$

where δ_F, p_F and t are measured in μm, bar and seconds respectively.

Oxygen strongly inhibits the process of fluorination of PS. Influence of O_2 concentration on A value (eq. 1) is shown in Figs. 7-8 and Table 1. Increase of an O_2 concentration results in a decrease of A value. O_2 concentration has the prevailing meaning, but some influence of fluorine partial pressure (when O_2 concentration is constant) was observed (Figure 8, Table 1). It was shown (see below) that treatment of polymers with F_2-O_2 mixtures (so-called oxyfluorination) resulted in a formation of C=O containing groups. It is known [19] that introduction of C=O groups in a polymer structure results in a substantial drop of permeability of different gases. It is reasonably to propose that introduction of C=O groups into fluorinated polymer structure decreases permeability of F_2 through fluorinated layer and hence results in a decrease of the fluorination rate.

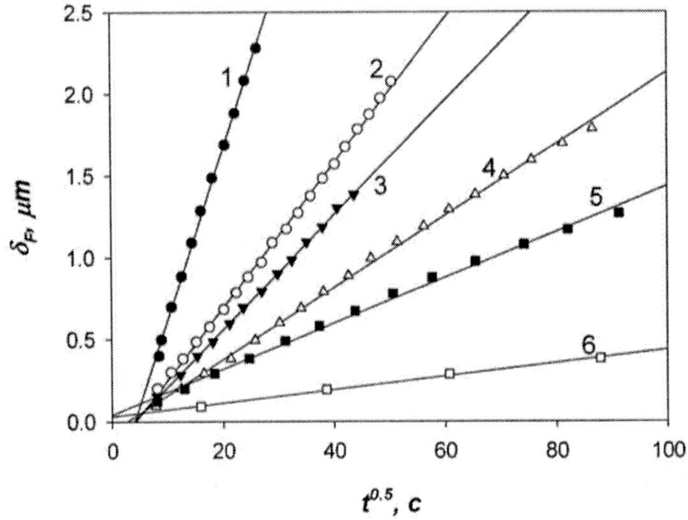

Figure 5. Thickness δ_F of fluorinated layer vs fluorination duration ***t***. Curves 1-6 correspond to the following compositions of fluorinating mixture (first figure in bracket correspond to the fluorine partial pressure in bars, second figure- to oxygen partial pressure in bars): (0.06; 0), (0.12; 0), (0.2; 0), (0.6; 0), (0.2; 0.05), (0.2; 0.2). Treatment temperature 293±1 K.

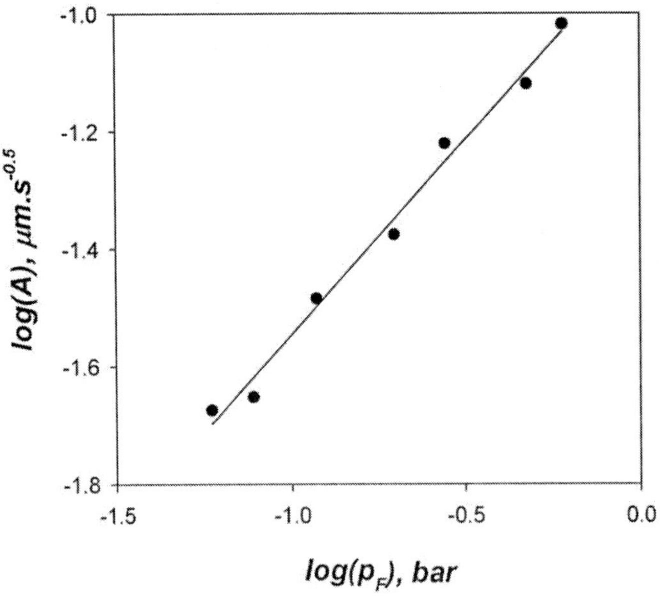

Figure 6. ***A*** value (form eq.) vs fluorine partial pressure p_F.

Table 1. Dependence of A, B and k (eq.) on oxygen concentration in fluorinating mixture. A_0 corresponds to the case of oxygen concentration equal to 0.

Oxygen concentration in fluorinating mixture, %	A/A_0	B, $\mu m \cdot bar^{-k} \cdot s^{-0.5}$	k
0	1.000	1.072	0.68
10	0.465	0.498	0.60
20	0.297	0.318	0.68
30	0.189	0.203	0.64
50	0.084	0.0906	0.61

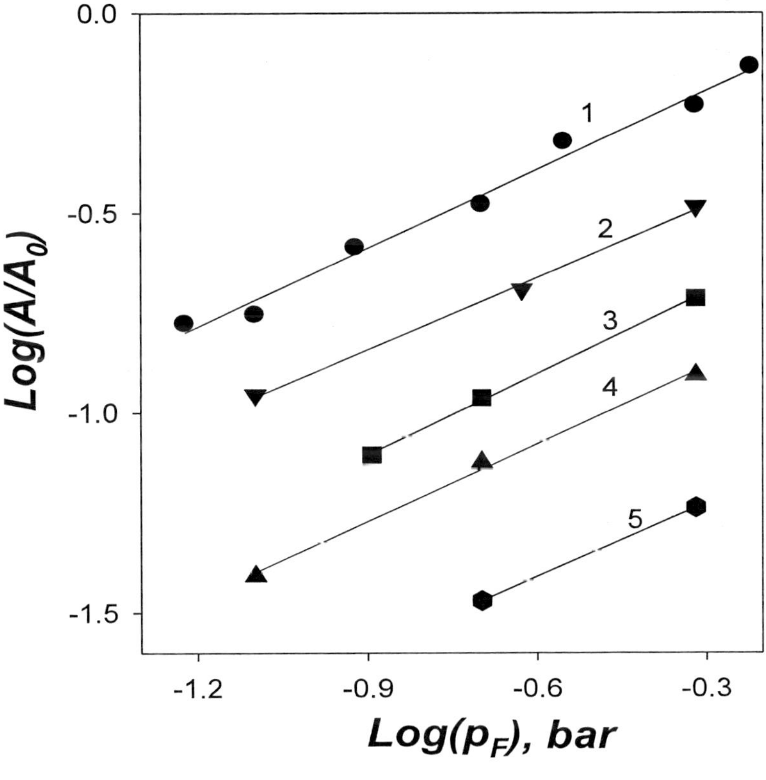

Figure 7. A/A_0 (see eq. ( )) vs fluorine partial pressure p_F in logarithmic scale. 1- treatment with undiluted fluorine. 2-5 correspond to fluorine concentration in fluorinating mixture 10, 20, 30 and 50 volume % respectively.

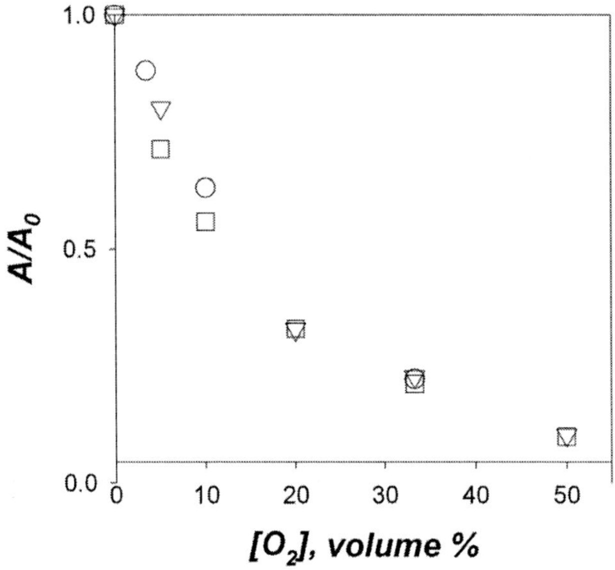

Figure 8. A/A_0 (A_0 is A value for the case of treatment with undiluted fluorine) vs oxygen percentage ***[O₂]*** in F_2-O_2 mixture. Circles, triangles and squares correspond to the following fluorine pressures: 0.08, 0.2 and 0.48 bar respectively.

Fluorination results in a substantial change of the polymer chemical composition of a PS film treated through all its thickness (Figure 9): all the bands corresponding to C-H and C-H_2 bonds and double conjugated C=C bonds disappear. Also several C=O-containing bands were detected in fluorinated PS: 1875 cm^{-1} (-COF) and 1790 cm^{-1} bands which are transformed into a broad 1750 cm^{-1} band by atmospheric moisture which also gives strong diffuse OH bands superimposed with bands of absorbed HF. Plausible explanation of this fact is as folows: -COF+H_2O⇒-COOH + HF [1, 6-13, 19]. The origin of C=O groups is due to admixture of oxygen in used fluorine and water absorbed on the reaction vessel walls. Absorption by C=O bands is also present in the near UV spectrum (band with maximum at 37000-38000 cm^{-1} or 260-270 nm, Figure 10). Thickness of the film used for Figure 10 was increased by a factor of 1.5 under fluorination (thickness of the virgin and fluorinated films can be calculated on the basis of interference features, because both virgin and fluorinated films were optically flat-parallel films). In the frames of another experiment thin PS film (10.2 μm in thickness) was cast from solution in chloroform, removed from a support and dried at elevated temperature. The film was placed inside a reaction vessel equipped with 2 optical windows fabricated from ZnSe. That material is stable to action of fluorine and transparent over 500-20,000 cm^{-1} spectral range. To remove HF emitted during the fluorination, NaF pellet was inserted inside the reaction vessel ($HF+NaF=NaHF_2$ (solid)). Then the reactor was evacuated and filed with fluorine at 0.1 bar pressure and 24 C temperature (filling duration did not exceed 10 seconds). The fluorine consumption at the end of fluorination did not exceed 7% so it can be regarded as constant. The reaction vessel was placed in the FTIR spectrometer and IR spectra were continuously monitored. The first 10 spectra (first spectrum was measured in 2 minutes after the injection of fluorine into the reaction vessel) were measured with 2 minutes intervals (30 seconds needed to collect 46 scans with 4 cm^{-1} resolution for each spectrum). Then amount of scans collected for each spectrum was increased to 100 (accumulation time 65 seconds). All the spectra were corrected with respect to absorption of ZnSe windows of the reaction vesse (Figure 11). The evolution of CH_2 bonds at 2924 and 2890 cm^{-1}, C-H bonds in a phenyl ring at 1493, 1451, 757 and 699 cm^{-1} and a very broad diffuse band over 870-1450 cm^{-1} (C-F, $C\text{-}F_2$ and $C\text{-}F_3$ bonds) [1, 6-13, 20] were continuously monitored. Computer simulation was used to separate overlapping bands and calculate the area $\boldsymbol{S}$ (cm^{-1}) under those bands in "Absorbance" mode, because the total amount of separate groups is proportional to the area of corresponding groups. Accuracy of the $\boldsymbol{A}$ value calculations can be evaluated as 5-10%. Separate bands were not extracted from a very broad diffuse band over 870-1450 cm^{-1} because those separate bands are rather broad and are not distinctly marked. Decrease of the relative amount $\boldsymbol{N/N_0}$ ($\boldsymbol{N}$ and $\boldsymbol{N_0}$ - amount of groups of a polymer chain at a moment $\boldsymbol{t}$ and in a virgin polymer respectively) and $\boldsymbol{S}$ value with square root of fluorination duration $\boldsymbol{t}$ is shown in Figure 11. Square root of $\boldsymbol{t}$ was used as a horizontal axe because the thickness $\boldsymbol{\delta_F}$ of a fluorinated layer is proportional just to $\boldsymbol{t^{0.5}}$. $\boldsymbol{S}$ value is proportional to $\boldsymbol{t^{0.5}}$ or, respectively, is proportional to the $\boldsymbol{\delta_F}$ value. It means that the total concentration of C-F, $C\text{-}F_2$ and $C\text{-}F_3$ bonds does not depend on the thickness of fluorinated layer. $\boldsymbol{N/N_0}$ values for all the C-H and $C\text{-}H_2$ bonds depend on the $\boldsymbol{t^{0.5}}$ value as follows:

$$N/N_0 = 1 - r \cdot t^{0.5} \quad (3)$$

where $\boldsymbol{r}$ coefficients for bands 1493, 1451, 757, 699 cm^{-1} and for a sum of 2924 cm^{-1} and 2890 cm^{-1} bands are equal to 0.0430, 0.0456, 0.0403, 0.0425 and 0.0411 respectively hence they practically coincide. The following conclusion can be made on the base of described experiments: amount of C-H and $C\text{-}H_2$ bonds is decreased simultaneously with increase of the total amount of C-F, $C-F_2$ and $C-F_3$ bonds and practically total transformation of virgin PS into fluorinates polymer takes place inside the boundary reaction zone.

As it was mentioned above according to the IR data no hydrogen was observed in a PS film treated through all its thickness. Chemical analysis indicated the following chemical formula of a monomeric unit of PS, fluorinated with undiluted fluorine: $C_8F_{12\pm0.5}$. The formula of untreated PS monomeric unit is C_8H_8 and fluorinated PS still containing conjugated bonds would be C_8F_8. If all the conjugated bonds are saturated with fluorine the content would be C_8F_{14}. It is known that under the action of fluorine both breakdown (disruption) and crosslinking take place [1, 21-23] hence the discrepancy between the abovementioned formulae (C_8F_{12}, C_8F_8 and C_8F_{14}) can be attributed to simultaneous breakdown and crosslinking of polymer chains and the number of crosslinks (or additional C-C bonds) per each monomeric unit is 2 units more than the number of scissions. The concentration of crosslinks is therefore equal to 1 per each monomeric unit of fluorinated PS. It is necessary to mention that our results do not agree with data for the chemical composition of fluorinated PS by other authors: C_8F_{14} [24] and $C_8F_{9.1}$ [25]. It may be due to a different treatment conditions.

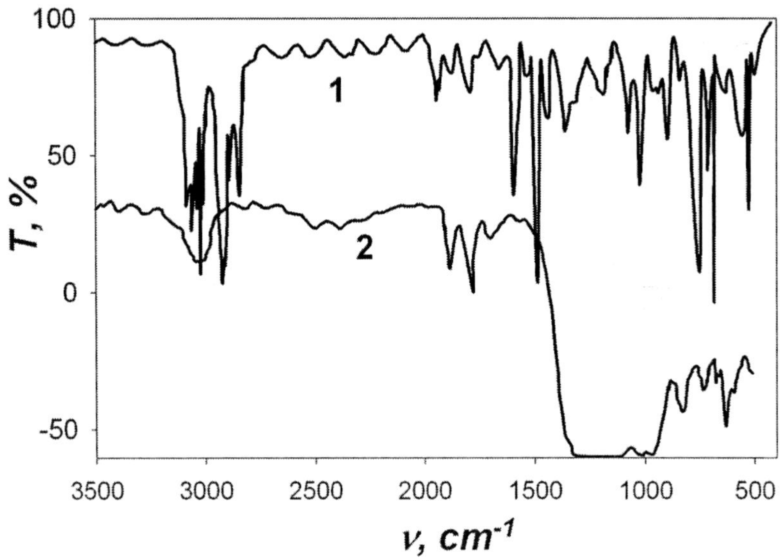

Figure 9. IR spectra of virgin (1) and fluorinated through all its thickness (2) PS film. Fluorine pressure 0.2 bar, temperature 295 C. Thickness of virgin and fluorinated films are equal to 20 and 30 μm respectively.

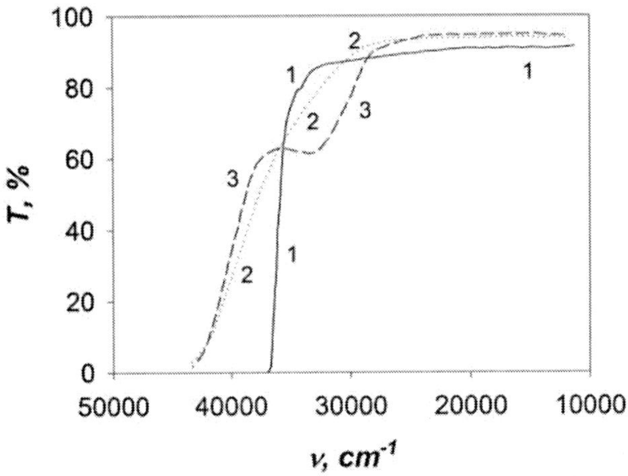

Figure 10. Transmittivity of the PS films: virgin, (1), treated with undiluted fluorine and fluorinated through all its thickness (2), and treated with $80\%F_2+20\%O_2$ mixture and fluorinated through all its thickness (3). The starting thickness of each film was equal to 25 μm. Treatment conditions: total pressure of fluorinating mixture was equal to 1.2 bar, fluorination duratyion 6 days, treatment temperature 294 K.) от от волнового числа ν.

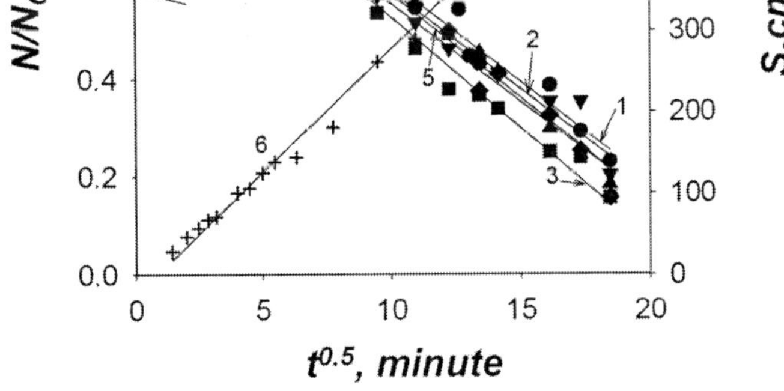

Figure 11. Relative amount N/N_0 (N and N_0 - amount of groups of a polymer chain at a moment t and in a virgin polymer respectively (curves 1-5) and S value (curve 6) on a square root of fluorination duration t for the case of treatment of PS film with fluorine (fluorine pressure 0.1 bar/ temperature 24 C). Figures 1-5 refers to the bands at 1493, 1451, 757, 699 cm^{-1} and for a sum of 2924 and 2890 cm^{-1} bands. 6- broad diffuse band over 870-1450 cm^{-1}.

Oxygen acts as an inhibitor of the direct fluorination. It means that the rate of the formation of the fluorinated layer decreases when concentration of oxygen is increased. The rate of the formation of the fluorinated layer is limited by permeability of F_2 through fluorinated layer to untreated one. Oxygen content in a fluorinating mixture strongly influences fluorine permeability through oxyfluorinated PS (Figures 5,7). To explain such a strong oxygen influence we used results of Salame [26] who had shown that the polymer permeability ***P*** is a function of the polymer chemical structure. The most marked features of oxyfluorinated PS are C=O 1875 cm^{-1} (-COF) and 1790 cm^{-1} bands [1, 6-13, 18-19]. Reaction with natmospheric moisture results in an origination of a broad 1750 cm^{-1} band and a strong diffuse OH bands superimposed with diffuse bands of absorbed HF. Plausible explanation of this fact is as follows: -COF+H_2O⇒-COOH + HF [1, 18-19]. Absorption by C=O bands is also present in the near UV spectrum (Figure 10). The R^{-1} value ($R= p_F/p_O$) increase leads to C=O bands intensity growth, moreover the 1790 cm^{-1} band intensity rises more rapidly than the 1875 cm^{-1}. The R value decrease results in a shift of the maximum of a very broad C-F_x band to lower frequencies. To verify the suggestion that C=O groups formation results in a fluorine permeability decrease we have measured C=O concentrations in oxyfluorinated PS.

Standard methods of C=O groups determination do not apply to oxyfluorinated PS due to its insolubility and atmospheric hydrolysis so we used IR spectroscopy. To avoid the influence of atmospheric moisture on IR spectra a reaction vessel with ZnSe optical windows was used and IR spectra were recorded in vacuum. We could not find quantitative data in the literature on absorption coefficients of carbonyl groups in fluorinated PS. To measure absorption coefficients we have used thin flat-parallel PMMA (polymethylmethacrylate) films (thickness 1 to 5 μm) cast onto ZnSe optical substrates (rather stable to fluorine action and transparent up to 500 cm^{-1}) and treated by fluorine. It should be noted that C=O groups are also present in an untreated PMMA (one group per monomer unit). Thicknesses of untreated and fluorinated layers were measured by the abovementioned interference spectroscopy method. The C=O groups extinction coefficients $\boldsymbol{\sigma}$ of oxyfluorinated PMMA [$\boldsymbol{\sigma(\nu)=\delta_F \cdot log(T_{background}/T(\nu))}$] versus frequency $\boldsymbol{\nu}$ are shown in Figure 12. We have made two assumptions in our analysis: (i) the concentrations of C=O groups in untreated and fluorinated PMMA are equal to each other being one carbonyl group per monomer unit; and (ii) the values of $\boldsymbol{\int\sigma(\nu)d\nu}$ for fluorinated PMMA and PS are coincident ($\boldsymbol{\int\sigma(\nu)d\nu}$ value is proportional to the C=O groups concentration). The difference between the $\int \sigma(\nu)d\nu$ values for untreated and fluorinated PMMA was less than 20%.

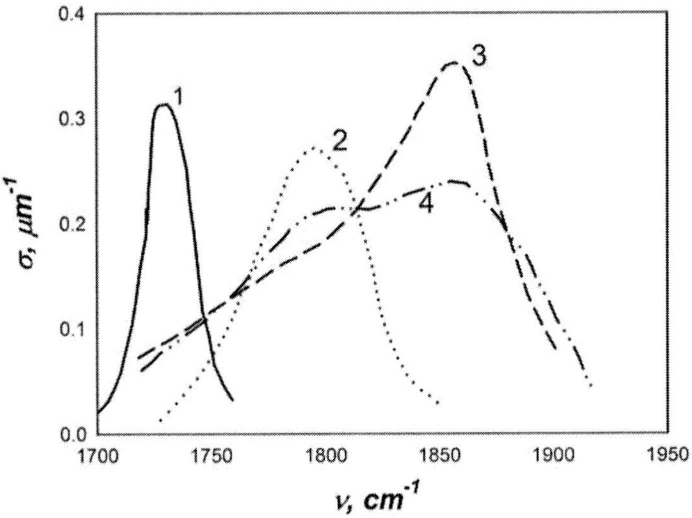

Figure 12. Extinction coefficient σ of C=O groups in untreated PMMA (1), PMMA treated with undiluted fluorine (2), PS treated with fluorine-oxygen mixture with R=4 (3) and PS treated with fluorine-oxygen mixture with R=0.33 vs wavenumber ν.

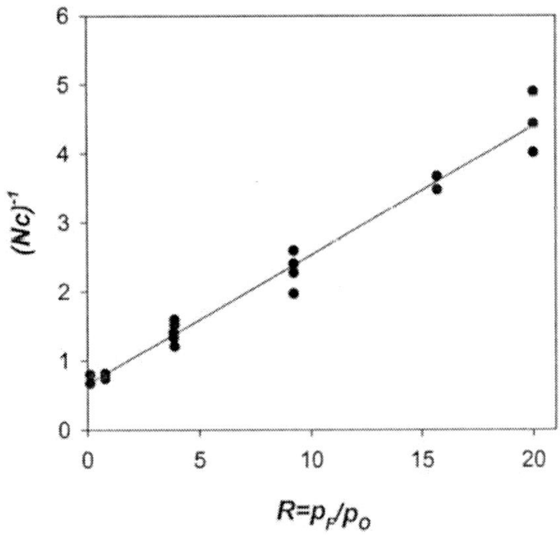

Figure 13. Number Nc of C=O groups per monomeric unit for oxyfluorinated PS vs R value.

The $\int\sigma(\nu)d\nu$ value was shown to be proportional to δ_F and hence the concentration N_c of C=O groups is fixed during the course of treatment. The concentration of C=O groups depends very weakly on p_F ($N_C \sim p_F^{-0.14}$) on treatment with a mixture of fixed R value.

The number N_C of C=O groups per monomer unit of fluorinated PS depends on the R value as follows (Figure 13):

$$N_C=1.53\cdot(1+0.29\cdot R)^{-1} \quad (4)$$

The value of N_C increases with oxygen concentration and reaches its maximum at ca. 1.5 when $R \to 0$.

According to [26] introduction of highly polar groups (such as hydroxyl OH and carbonyl C=O) into polymer structures results in a large decrease of permeability. For example, on comparing two polymers, polyethylene $(-CH_2-CH_2-)_n$ and polyvinylalcohol $(-CH_2-CHOH-)_n$, the difference between permeability values for these polymers is found to exceed six orders of magnitude due only to one hydroxyl OH group per monomer unit. So the observed strong C=O group influence on the permeability is not surprising.

Not only formation of C=O groups can result in permeability decrease. The same effect can also arise due to a formation of crosslinks such as "oxygen bridges" -C-O-C-. Unfortunately standard methods of crosslinking determination (e.g. swelling method) cannot be used. IR spectroscopy cannot be applied to detect -C-O-C- groups because the absorption bands of these groups are usually situated near the 1100-1300 cm^{-1} [20] and are masked by a very strong absorption of carbon-fluorine bonds.

On the base of abovementioned results the following simplest model of the kinetics of fluorinated layer formation can be proposed. At first it is necessary to underline the principle difference between the diffusion of the gas in a polymer matter for two case: (a) gas does not react with a polymer and (b) the reaction of a gas with polymer takes place. The gas concentration profile is shown in the Figure 14. For the case of diffusion without reaction (left picture in the Figure 14) the concentration of a gas is described by a well known diffusion equation placed at that picture. But for the case when a gas (namely F_2) penetrates through fluorinated layer ($\delta(t) \equiv \delta_F$ in thickness, right picture in the Figure 14; vertical line correspond to the thickness of fluorinated layer δ_F) the concentration profile (indicated in the picture) strongly differs from the previous case.

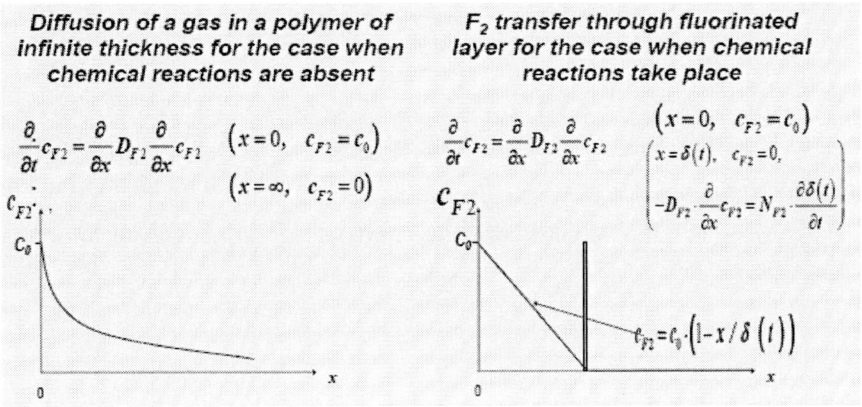

Figure 14. Gas profile concentration for the case when gas does not react with a polymer (left picture) and the reaction of a gas with polymer takes place (right picture).

As it was shown above, the treated film consists of layers of fluorinated polymer and of unmodified one (Figure 2). There is a very thin transition (boundary) layer in between this two layers ($\boldsymbol{\delta_b}$ in thickness; $\boldsymbol{\delta_b}<<\boldsymbol{\delta_F}$), where the main chemical transformations take place. In this case the rate of formation of the fluorinated layer is limited by the permeability $\overline{\boldsymbol{P}}_{F2}$ of F_2 through the fluorinated layer to the transition zone. Assuming that the $\overline{\boldsymbol{P}}_{F2}$ value remains unchanged during fluorination and fluorine partial pressure is maintained constant, the balance between the amount of F_2 consumed in the reaction zone and the amount of F_2 passed through the fluorinated layer can be expressed by the following relation:

$$N_{F2}\cdot\frac{d\delta_F}{dt}=D\cdot\frac{dC_F}{dx} \tag{5}$$

where $\boldsymbol{D}$ is the diffusion coefficient of fluorine in the fluorinated polymer; $\boldsymbol{C}_F$ is the concentration of fluorine in the fluorinated polymer, axe x is directed from the surface to the polymer bulk perpendicularly to the surface (x=0 at the film surface). Experimentally measured coefficient N_{F2} is equal to the amount of F_2 (in cm^3 (STP)) necessary to form 1 cm^3 of the fluorinated polymer divided by 1 cm^3 of the fluorinated polymer. Assuming that $\delta_b<<\delta_F$, $c_F(x=\delta_F)<<c_F(x=0)$ and properties of the fluorinated layer are not changed along the axis $\boldsymbol{x}$ we have

$$N_{F2} \cdot \frac{d\delta_F}{dt} = D \cdot \frac{C_F(0)}{\delta_F} = D \cdot S \cdot \frac{p_F}{\delta_F} \equiv \overline{P}_{F2} \cdot \frac{p_F}{\delta_F} \quad (6)$$

where S is the solubility coefficient of fluorine in fluorinated polymer ($c_F(x=0)=S \cdot p_F$ and $\overline{P}_{F2}=D \cdot S$). Under steady-state conditions of fluorination and assuming that the partial fluorine pressure and $\overline{P}$ are not changed during fluorination

$$\delta_F = \left[\frac{2 \cdot \overline{P}_{F2} \cdot p_F \cdot t}{N_{F2}}\right]^{0.5} = A \cdot t^{0.5} \quad (7)$$

To measure the fluorinated PS density the following procedure was applied. PS film was cast onto an optical support and a flat-parallel film was selected. The thickness of the film was calculated from its visible spectra as it was described above. The weight and area of the film were also measured. Then the film was fluorinated through all its thickness. The thickness of fluorinated film was measured also from the visible spectra and its weight and area were measured (area of the film was not changed but the thickness was increased by a factor of 1.5 with respect to the starting film). From these data the density of virgin and fluorinated PS were measured as 1.05 and 2.05 g.cm^{-3} respectively. Hence fluorination results in a significant, almost double, increase of the PS density.

At present time majority of researches [1, 27-32] consider process of the direct fluorination of polymers as a radical-chain one. But such an affirmation was based only on the fact that long-life radicals were detected in the fluorine treated polymers, but no radicals taking part in the processes of chain initiation and propagation were detected except research carried out by Kharitonov et al. [7-13, 33-34]. Authors of [1, 21, 23, 35-36] detected long-lived radicals in fluorinated (at room temperature and up to 55 C) PS and other polymers. ESR spectra represented a mixture of peroxy RO_2 radicals and nonindentified very broad component, which was tentatively assigned to metal fluorides with altering valency [21]. But assignment of those radicals to metal fluorides with altering valency was refuted by the authors of [33-34, 37]. The concentration of radicals was varied over 10^{16}-10^{19} radicals per 1 gram of a total mass of a polymer. But it is necessary to underline that the above value does not represent a real concentration of radicals inside fluorinated layer, because polymers were not fluorinated through all their thickness and the thickness of a fluorinated layer was not measured.

When PS is treated with undiluted fluorine (concentration of O_2 admixture is less than 0.1%), ESR spectrum of fluorinated PS consists of superimposed asymmetric peroxy $RO_2^{\bullet}$ singlet and a very broad component possessing superfine structure (1, Figure 15; measured at 77 K) [33-34], which can be attributed to a fluoro-containing radical. The total spectrum width is about 16.5 mT. At a room temperature (2, Figure 15) *g*-factor is averaged due to unrestricted rotation of peroxy $RO_2^{\bullet}$ radical. Low-field component of G-factor, corresponding to $g_{||}$ component of *g*-factor, becomes unresolved. Increase of oxygen concentration in O_2/F_2 mixture results in increase of relative (with respect to the total amount of radicals) concentration of peroxy RO_2 radicals (3, Figure 15). To clarify the nature of a broad component of the spectrum, samples of fluorinated PS were gradually heated up to 550 K. Increase of a temperature resulted in a gradual decrease of absorption of both peroxy radical and broad component and at 550 K practically all the radicals terminated. Hence a broad spectrum has a radical nature rather than admixtures of metal fluorides [21], because spectra of the latter species should not depend on temperature.

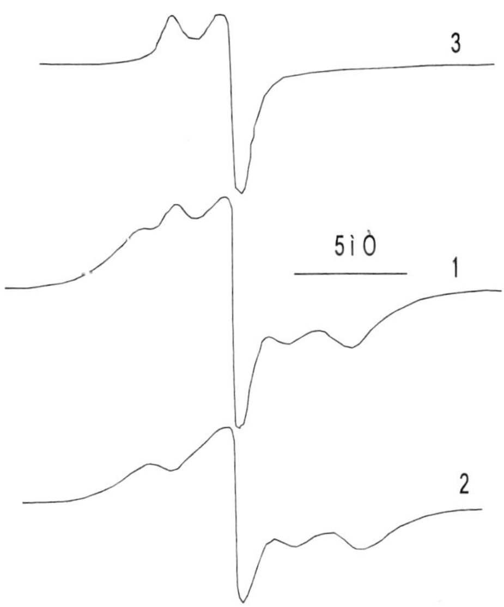

Figure 15. ESR spectra of PS treated at 295 K with 80%F_2+20%O_2 mixture (1) and undiluted fluorine (2 and 3). Spectra 1 and 2 were measured at 77 K, spectrum 3- at 300 K.

To measure kinetics of accumulation of radicals during fluorination, PS film was placed inside a Teflon vial, which was inserted inside the resonator of ESR spectrometer [33]. Then mixture of 80%F_2+20%O_2 at total pressure 0.24 bar at 22 C temperature was inserted into the vial and ESR spectra were monitored "*in situ*". The dependence of peroxy $RO_2^{\bullet}$ radicals amount on the thickness $\boldsymbol{\delta}_F$ of fluorinated layer is shown in the Figure 16. Thickness of $\boldsymbol{\delta}_F$ on fluorination duration was calculated on the basis of previously obtained data (see above). Radicals originate just after insertion of fluorinating mixture into the vial, but at an initial stage there is no linear dependence of radicals amount on $\boldsymbol{\delta}_F$ value. Some time (or respectively formation of fluorinated layer 0.3-0.4 μm in thickness) is needed to reach the linear dependence.

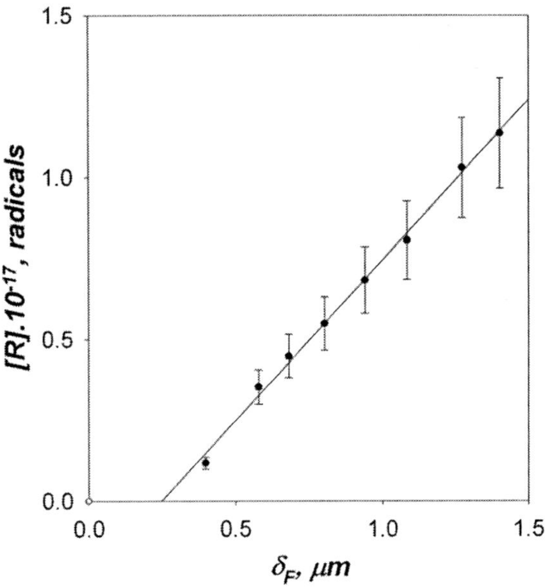

Figure 16. Dependence of the total amount of radicals [R] in PS film treated with 80%F2+20%O2 mixture at a total pressure 0.24 bar at 220C on the fluorinated layer thickness □F.

Conclusion

1. The direct fluorination of polystyrene is a diffusion-controlled process. This indicates that the rate of formation of a fluorinated layer is limited by the rate of penetration of fluorine F_2 through the

fluorinated polymer layer into the untreated layer. The fluorinated and untreated polymer layers are separated by a very thin transient boundary layer where the majority of the chemical reaction takes place.

2. The direct fluorination of polystyrene results in a practically total disruption of C-H bonds and saturation of double (conjugated) bonds with fluorine followed by a formation of C-F, $C\text{-}F_2$ и $C\text{-}F_3$. Also C=O-containing groups are formed.
3. The dependence of the fluorinated layer thickness $\boldsymbol{\delta_F}$ on fluorination duration $\boldsymbol{t}$ and fluorine partial pressure $\boldsymbol{p_F}$ is described by the following formula: $\boldsymbol{\delta_F = 0.14 \cdot (p_F)^{0.68} \cdot t^{0.5}}$ at temperature 293±1 K, where $\boldsymbol{\delta_F}$, $\boldsymbol{p_F}$ and $\boldsymbol{t}$ are measured in μm, bar and seconds respectively.
4. . The rate of formation of a fluorinated layer increases with fluorine partial pressure is not practically affected by the presence of N_2, He, Ar and CO_2 in the fluorinating mixture. O_2 inhibits the fluorination rate.
5. Fluorination results in a decrease of the refraction index of polymers.
6. Density of fluorinated layer substantially exceeds density of the virgin polymer.
7. The fluorination of PS is a chain-radical process. Peroxy RO_2 and fluororadicals are present in the fluorinated polymers in a relatively large concentration. Those radicals are long-life radicals: their amount is decreased by a factor of ~15 hours.

REFERENCES

[1] Lagow; R.J; Margrave, J.L. Direct fluorination: a "new" approach to fluorine Chemistry. *Progr. in Inorg. Chem.* 1979, 26, 162-210.

[2] Jagur-Grodzinski, J. Modification of polymers under heterogeneous conditions. *Progr. in Polymer Sci.* 1992, 17, 361-415.

[3] Anand, M.; Hobbs, J.P.; Brass, I.J. Surface fluorination of polymers. In *Organofluorine Chemistry: Principles and Commercial Applications,* (eds) Banks, R.E.; Smart, B.E.; Tatlow, J.C. Plenum Press: New York, 1994; pp.469–481.

[4] Carstens, P.A.B.; Marais, S.A.; Thompson, C.J. Improved and novel surface fluorinated products. *J. Fluorine Chem.* 2000, 104, 97-107.

[5] Taege, R.; Ferrier G. *European Coatings. J.* 2006, Nos.5-6, 36-40.

[6] Hruska, Z. Industrial oxyfluorination of polymer web materials. *Intern. Conf. "Fluorine in Coatings-IV"*, Brussels, Belgium, 5-7 March 2001.- Conference papers.-Paper 34.

[7] Kharitonov, A.P. Practical applications of the direct fluorination of polymers. *J. Fluorine Chem.* 2000, 103, 123-127.

[8] Kharitonov, A.P.; Taege, R.; Ferrier, G.; Teplyakov, V.V.; Syrtsova, D.A.; Koops, ;G.H. Direct fluorination - useful tool to enhance commercial properties of polymer articles. *J. Fluorine Chem.* 2005, 126, 251-263.

[9] A.P. Kharitonov. Chapter 2: Direct fluorination of polymers—from fundamental research to industrial applications. In: *Fluorine Chemistry Research Advances*, Ed. Ira V. Gardiner. Nova Science Publishers, Inc., N.Y., 2007. P.35-103.

[10] Tressaud, E. Durand, C. Labruge`re, A.P. Kharitonov, L.N. Kharitonova. Modification of surface properties of carbon-based and polymeric materials through fluorination routes: From fundamental research to industrial applications. *J. Fluorine Chem.* 2007, 128, 378–391.

[11] P. Kharitonov. *Direct fluorination of polymers*. Nova Science Publishers Inc. N.Y., 2008.

[12] A.P. Kharitonov, L.N. Kharitonova. Surface modification of polymers by direct fluorination: A convenient approach to improve commercial properties of polymeric articles. *Pure and Appl. Chem*. 2009, 81, 451-471.

[13] A.P. Kharitonov, B.A. Loginov. Direct fluorination of polymer final products: From fundamental study to practical application. *Russian J. of General Chem*. 2009, 79, 635-641.

[14] Kharitonov, A.P.; Moskvin, Yu.L.; Kolpakov, G.A. Application of the interference spectroscopy method to study the kinetics of chemical reactions in optically transparent films. *Polymer Science USSR,* 1985, 27, 739-743.

[15] Kharitonov, A.P.; Moskvin, Yu.L.; Kharitonova, L.N.; Kotenko, A.A.; Tulskii, M.N. Use of interference methods for study of the fluorination kinetics of homogeneous and composite polymeric membranes. *Polymer Science, Ser.B,* 1995, 37, 307-310.

[16] Kharitonov, A.P.; Moskvin, Yu.L.; Kolpakov, G.A. The direct fluorination of polyethylene terephthalate films. *Sov. J. Chem. Phys.* 1987, 4, 877-885.

[17] Shimada J.; Hoshino, M. Surface fluorination of transparent polymer films. *J. Appl. Polymer Sci.* 1975, 19, 1439-1448.

[18] Kharitonov, A.P.; Moskvin, Yu.L. Direct fluorination of polystyrene films. *J. Fluorine Chem.* 1998, 91, 87-93.
[19] Kharitonov, A.P.; Moskvin, Yu.L.; Teplyakov, V.V.; Le Roux, J.D. Direct fluorination of poly(vinyl trimethylsilane) and poly(phenylene oxide). *J. Fluorine Chem.* 1999, 93, 129-137.
[20] Socrates, G. *Infrared Characteristic Group Frequences.* 2nd edition, Chichester; N.Y.;Brisbane; Toronto: John Wiley & Sons, 1994.
[21] Florin, R.E. Electron spin resonance spectra of polymers during fluorination. *J. Fluorine Chem.* 1979, 14, 253-262.
[22] Brown, D.W.; Florin R.E.; Wall, L.A. *Appl. Polymer Symp.* 1973, No.22, 169.
[23] Toy M.S.; Newman, J.M. Polyperfluorobutadiene. II. Fractionation and crosslinking. *J. Polymer Sci. A-1,* 1969, 7, 2333-2340.
[24] Lagow, R.J.; Margrave, J.L. The controlled reaction of hydrocarbon polymers with elemental fluorine. *J. Polymer Sci., Pol. Lett. Ed.* 1974, 12, 177-184.
[25] Shimada J.; Hoshino, M. Surface fluorination of transparent polymer films. *J. Appl. Polymer Sci.* 1975, 19, 1439-1448.
[26] Salame, M. *Polymer Preprints.* 1964, 8, 137.
[27] Moorehead, A.; Margrave, J.L. The direct fluorination of plastics at room temperature. *2nd International Conference 'Fluorine in coatings'.* Salford, England, 28-30 September 1994.-Conference papers.-Paper N.15.
[28] Hara, N.; Fukumoto, H.; Watanabe, M. In-situ kinetic study on direct fluorination of thin polyethylene films with QCM. *Bull. Chem. Soc. Jpn.* 1995, 68, 1232-1238
[29] Sanderson, R.D.; du Toit, F.J.; Carstens, P.A.B.; Wagener, J.B. Fluorination rates of polyolefins as a function of structure and gas atmosphere. *J. Thermal Analysis.* 1994, 41, 563-581.
[30] Nazarov, V.G. Composition and dimensions of the surface and transient layers in modified polymers. *Vysocomol. Soedineniya, ser.B* (Polymer Science Ser.B) (in Russian). 1999, 41, 734-738.
[31] Friedrich, J.; Kuhn, G.; Schulz, U.; Jansen, K.; Moller, B.; Fischer, S. *Vakuum in Forschung und Praxis.* 2002, 14, 285-290.
[32] Clark, D.T.; Feast, W.J.; Musgrave, W.K.S.; Ritchie, J. Applications of ESCA to polymer Chemistry. Part VI. Surface fluorination of polyethylene. Application of ESCA to the examination of structure as a function of depth. *J. Polymer Sci.: Polymer Chem. Ed.* 1975, 13, 857-890.

[33] Kolpakov, G.A.; Kuzina, S.I.; Kharitonov, A.P.; Moskvin, Yu.L.; Mikhailov, A.I. Free radical accumulation during direct fluorination of polystyrene. *Sov. J. Chem. Phys.* 1992, 9, 2283-2287.

[34] S Kuzina, S.I.; Kharitonov, A.P.; Moskvin, Yu.L.; Mikhailov, A.I. Formation of free radicals in the low-temperature fluorination of polymers. *Russian Chem. Bull.* 1996, 45, 1623-1627.

[35] Hayes, L.J.; Dixon, D.D. *J. Fluorine Chem.* 1977, 10, 1-6.

[36] Florin, R.E.; Wall, L.A. Radicals detected by electron spin resonance during fluorination of polymers. *J. Chemical Phys.* 1972, 57, 1791-1792.

[37] Kuzina, S.I.; Mikhailov, A.I.; Gol'danskii, V.I. Spontaneous formation of free radicals at low temperatures. *Eur. Polym. J.* 1995, 31, 513-519.

In: Polystyrene: Properties, Performance… ISBN: 978-1-61209-117-4
Editor: James E. Gray

Chapter 5

POLYSTYRENE: PROPERTIES AND IT'S APPLICATIONS IN SENSING PLATFORMS

***Gaurav Chatterjee*[1] *and Shalini Prasad**[*2]
[1]Department of Electrical, Computer and Energy Engineering, Arizona State University, Tempe, AZ, USA
[2]Department of Electrical Engineering and Computer Science, Wichita State University, Wichita, KS, USA

ABSTRACT

Polymeric materials have gained a wide theoretical interest and practical application in sensor technology. Polymers offer a lot of advantages for sensor technologies: they are relatively low cost materials, their fabrication techniques are quite simple (there is no need for special clean-room and/or high temperature processes), they can be deposited on various types of substrates and the wide choice of their molecular structure and the possibility to build in side-chains, charged or neutral particles, and even grains of specific behavior into the bulk material or on its surface region, enables films to be produced with various physical and chemical properties. These properties can be adapted for sensing behavior.

Out of the various polymers currently available, polystyrene, because of its preferable physical, optical and electrical properties, has generated a

---

* Corresponding Author: Shalini Prasad, Associate Professor, EECS, Bomhoff Distinguished Professor of Bioengineering, PO Box 44, Wichita State University, Wichita, KS-67207

considerable amount of interest in the sensor community. Polystyrene has found applications more specifically in the biological, chemical and optical sensor areas.

The scope of this review chapter is to demonstrate the applications of polystyrene into various sensing platforms. The various physical and chemical forms in which polystyrene is used are elaborated. The various detection methods (electrical, optical, mass based) are also elaborated. Finally, the present status and perspectives as well as the advantages of the specific polystyrene based sensors are summarized.

1. INTRODUCTION

During the last 25 years, global research and development (R&D) on the field of sensors has expanded exponentially in terms of financial investment, the published literature, and the number of active researchers. It is well known that the function of a sensor is to provide information on our physical, chemical and biological environment. Legislation has fostered a huge demand for the sensors necessary in environmental monitoring, e.g. monitoring toxic gases and vapors in the workplace or contaminants in natural waters by industrial effluents and runoff from agriculture fields. Thus, a near revolution is apparent in sensor research, giving birth to a large number of sensor devices for medical and environmental technology. A chemical sensor furnishes information about its environment and consists of a physical transducer and a chemically selective layer. A biosensor contains a biological entity such as enzyme, antibody, bacteria, tissue, etc. as recognition agent, whereas a chemical sensor does not contain these agents. Sensor devices have been made from classical semiconductors, solid electrolytes, insulators, metals and catalytic materials. Since the chemical and physical properties of polymers may be tailored by the chemist for particular needs, they gained importance in the construction of sensor devices. Although a majority of polymers are unable to conduct electricity, their insulating properties are utilized in the electronic industry. A survey of the literature reveals that polymers also acquired a major position as materials in various sensor devices among other materials (Table 1). Either an intrinsically conducting polymer is being used as a coating or encapsulating material on an electrode surface, or non-conducting a polymer is being used for immobilization of specific receptor agents on the sensor device [1].

Table 1. Summary of the status, advantages and disadvantages of polymer based sensors [2]

Sensor application	Avantages compared to inorganic types	Disadvantages compared to inorganic types	Present status
PTC thermistors	Sharp PTC effect in a wide temperature range	Limited operation temperature range	Available on the market
Pressure sensors	Very high sensitivity, shock resistance	Stability problems	Research and development
Tactile sensors	Flexibility, the best simulation of human fingers	–	Research and development
Acoustic sensors	High sensitivity, shock resistance	Limited temperature range	Available on the market
Infra-red radiation sensors	High thermal resistivity, arrays with low cross-talk	Low sensitivity	Available on the market, arrays under development
Humidity sensors	High sensitivity, integration possibility	Stability problems	Available on the market
Gas sensors	High selectivity, high sensitivity, room temperature operation	Long-term drift	A few types are available on the market, others are under development
Ion-selective sensors	High selectivity, a wide choice of ionophores	Short lifetime	Electrodes available on the market, integrated types under development
Biosensors	Biocompatibility, easy immobilization of enzymes or other bioactive compounds	–	Research and development, some types on the market

2. Sensor Structures with Polymers as Active Elements

The typical device structures, that can be used to measure the changes in the properties of polymer films, can be categorized into the following groups[2]:

- impedance type sensors

Follow the sensing phenomena with capacitance and/or with resistance changes.

- semiconductor device based types

These are more complicated; the sensing effect can change the most important characteristics and/or parameter of the devices, for example shift the diode characteristics, alter the threshold voltage of a field effect transistor (FET), etc.

- resonant sensors

The shift of the resonance frequency can be measured as a function of the quantity to be measured due to the changes of mass or to changes in wave propagation properties. According to the types of the waves, they can be grouped into bulk acoustic, surface acoustic and flexural plate wave (BAW, SAW, FPW) devices.

- electrochemical cells

These are widely used in chemical sensors; the electrode potential, the cell current and/or the cell resistance can be measured as a function of the analyte, which is in close contact with the sensing polymer film; thus, potentiometric, amperometric, and conductimetric devices can be distinguished.

- calorimetric sensors

These are based on the measurement of a temperature changes caused by physical or chemical process effects from the environment.

- Fiber optic sensors

These represent a new generation of sensors; they are based on the changes of the light propagation, absorption, and/or emission properties in the sensing polymer films and give an optical output signal. The possible structures are: optrode-style, core-based, coating-based, interferometric, and other type fibre-optic sensors.

Sensing Effects in Polymers [2]

Sensing effects are those physical and chemical phenomena that are the bases of the sensor's operation i.e. provide an electrical or optical signal as a function of the quantity to be measured. It is a difficult task to give a proper categorization of the sensing effects due to the wide variety of material structures and physical/chemical interaction phenomena that can be used to fabricate sensors based on sensing polymer films. The sensing effects in polymers can be described using the following grouping of materials: dielectrics, conductive or other composites, electrolytes, electro-conducting conjugated polymers, sorbent materials, ion-exchange membranes, perm-selective membranes, membranes with specific recognition sites, optically sensitive polymers.

3. POLYSTYRENE AS A POLYMER

Polystyrene, commonly known as 'Styrofoam' is one of the most widely used type of plastics. Polystyrene (PS) is a clear, colorless polymer used extensively for low-cost applications. Polystyrene is actually an aromatic polymer that is made from the monomer styrene. It is a long hydrocarbon chain that has a phenyl group attached to every carbon atom. Styrene is an aromatic monomer, commercially manufactured from petroleum. Polystyrene is a vinyl polymer, manufactured from the styrene monomer by free radical vinyl polymerization.Polystyrene is a rigid, transparent thermoplastic, which is present in solid or glassy state at normal temperature. But, when heated above its glass transition temperature, it turns into a form that flows and can be easily used for molding and extrusion. It becomes solid again when it cools off. This property of polystyrene is used for casting it into molds with fine detail. Pure polystyrene polymer is colorless and hard with limited flexibility. It can be transparent or can be made to take on different colors. It is available commercially in both pellet and sheet form. The most serious deficiencies are low impact strength, poor weatherability and poor chemical resistance. Numerous modified grades which seek to correct these shortcomings are commercially available.

Properties of Polystyrene

The unique physical and chemical properties of polystyrene are responsible for its use in a wide range of applications. Polystyrene is hard and brittle and has a density of 1.050 g/cm^3. It is represented by the chemical formula, C_8H_8. It is made up of three chemical elements, carbon, hydrogen and oxygen. Most of the polystyrene properties are as a result of the unique properties of carbon. It is highly flammable and burns with an orange yellow flame, giving off soot, as a characteristic of all aromatic hydrocarbons. Polystyrene, on oxidation, produces only carbon dioxide and water vapor.

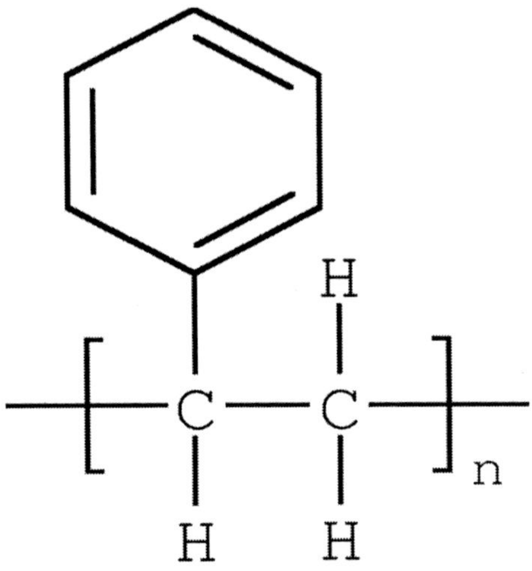

Figure 1. Chemical structure of polystyrene.

The physical properties of polystyrene are given below:

- Density - 1.05 g/cc
- Dielectric constant - 2.4 to 2.7
- Thermal conductivity - 0.08 W/(m.K)
- Young's modulus - 3000 to 3600 Mpa
- Tensile strength - 46 to 60 Mpa
- Melting point - 240 °C
- Water absorption - 0.03 to 0.1

Polystyrene is chemically nonreactive and hence, used to make containers for other chemicals, solvents and even food items. The transformation of carbon-carbon double bonds into less reactive single bonds in polystyrene is the main reason for its chemical stability. The chemical stability, mechanical strength and the fact that PS can be easily modified to alter its properties make it a very attractive candidate for sensor applications.

4. Applications of Polystyrene in Sensor Platforms:

Bio-Sensors

The unprecedented interest in the development and exploitation of analytical devices for detection, quantification and monitoring of specific chemical species has led to the emergence of biosensors. The estimation of metabolites such as glucose, urea, cholesterol and lactate in whole blood is of central importance in clinical diagnostics. Biosensors represent a new trend emerging in the diagnostic technology. A biosensor is a device having a biological sensing element either intimately connected to or integrated within a transducer. The aim is to produce a digital electronic signal, which is proportional to the concentration of a specific chemical or set of chemicals. Biosensor instruments are specific, rapid, simple to operate, can be easily fabricated with minimal sample pretreatment involved. The apparently alien marriage of two contrasting disciplines combines the specificity and sensitivity of biological systems with the computing power of microprocessor [3].

Bio-components, which function as biochemical transducers can be enzymes, tissues, bacteria, yeast, antibodies/ antigens, liposomes, organelles. Within a biosensor the recognition biomolecule incorporated possesses an exquisite level of selectivity but is vulnerable to extreme conditions such as temperature, pH and ionic strength. Most of the biological molecules such as enzymes, receptors, antibodies, cells etc. have very short lifetime in solution phase. Thus they have to be fixed in a suitable matrix. The immobilization of the biological component against the environmental conditions results in decreased enzyme activity. The activity of immobilized molecules depends upon surface area, porosity, hydrophilic character of immobilizing matrix, reaction conditions and the methodology chosen for immobilization. A number of techniques such as physical adsorption, cross-linking, gel entrapment, covalent coupling etc. have been used to immobilize biological molecules in carrier materials (Table 2).

The industry standard for bio-molecule detection is using the ELISA technique. Immunoreagents attached to a solid phase are extensively used in immunoassays (radioimmunoassay, ELISA). Solid-phase reagent quality depends on several parameters such as the quality of the solid backbone and the level of adsorption/desorption during the assay. Due to their highly reproducible adsorption characteristics and experimental simplicity, polystyrene microtiter plates have become the most commonly used solid phases for immunoassays [4]. A very comprehensive adsorption-desorption study was carried out by Nieto et al[4] which established various affinity constants of the bio molecules with the solid phase.

Table 2. Conventional immobilization procedures

Method	Advantages	Disadvantages	Example
Physical adsorption	No modification of biocatalyst. Matrix can be regenerated. Low cost	Binding forces are susceptible to change in pH, temperature and ionic strength	Adsorption of glucose oxidase on conducting polymers for glucose detection
Entrapment	Only physical confinement of biocatalyst near transducer. Low cost	High diffusion barrier	Entrapment of urease and glutamate dehydrogenase in polypyrrole/polyvinyl sulphonate films for urea detection
Cross-linking	Loss of biocatalyst is minimum. Moderate cost	Harsh treatment of biocatalyst by toxic chemicals	Glutaraldehyde mediated linking of lactate dehydrogenase for lactate estimation
Covalent bonding	Low diffusional resistance Stable under adverse conditions	Harsh treatment by toxic chemicals. Matrix not regenerable	GOD binding via poly(*o*-amino benzoic acid) for glucose detection

The fact that antibodies are readily adsorbed to polystyrene films was exploited by Brizzolara et al [5] to pattern multiple antibodies onto a single polystyrene substrate. The idea was borrowed from the standard photolithography technique used in solid state to pattern silicon. The surface is initially coated with a material resistant to antibody adsorption, and the coating is selectively removed using ion beam sputtering or mechanical etching to expose the areas where antibody adsorption is desired. Then those areas are exposed to the antibody. Patterning multiple antibodies on a single substrate enables the platform to perform multiplexed biosensing. This has a potential in point of care diagnostic situations.

The adsorbing property of polystyrene was employed by Bouafsoun et al [6] to electrically probe the endothelial cell behavior at a thiol functionalized gold surface using faradaic electrical impedance spectroscopy (EIS). By virtue of its unique location in the vessel wall, the endothelium constitutes an essential interface. It is important to characterize endothelial cell adhesion to be able to predict and control the outcome of the interaction between artificial surfaces and living cells, so that biomedical devices of antithrombogenic vascular grafts can be developed.

Cross-linking of polystyrene introduces a capillary membrane like function to the normal polystyrene. Lightly cross-linked gel type polystyrene (GPS) has been most widely used due to its common availability and inexpensive cost. GPS beads which are functionalized with chloromethyl-, aminomethyl-, and a variety of linkers are commercially available from a variety of sources. A prominent characteristic of GPS beads is their ability to absorb large relative volumes of certain organic solvents (swelling). This swelling causes a phase change of the bead from a solid to a solvent-swollen gel, and therefore, the reactive sites are accessed by diffusion of reactants through a solvent-swollen gel network. In solvents, which swell the polymer well, the gel network consists of mostly solvent with only a small fraction of the total mass being polymer backbone. This allows relatively rapid diffusional access of reagents to reactive sites within the swollen bead. In solvents, which do not swell the polymer, the cross-linked network does not expand and the diffusion of reagents into the interior of the bead is impeded. This property of cross linked polystyrene (CLPS) has been used as a carrier mechanism for bioreagents [7]. Poly(p-maleimidostyrene) (PMS) was identified as a novel polymer having densely pendent maleimido groups [8]. Preparation of a polymeric material with stable biomolecule, such as an enzyme, immobilization characteristics, even after ultrasonic cleaning, is very important, not only for the fabrication of biosensors and bioreactors, but also for the development of the enzyme linked immunosorbent assay (ELISA) in the clinical field. PMS, which has an excellent capacity to immobilize biomolecules, should therefore be coated onto a carrier matrix with a polystyrene structure. This would be beneficial owing to the satisfactory affinity of PMS for polystyrene, owing to its polystyrene framework. A urea sensor was fabricated by combining an ammonia electrode with urease-immobilized PMS/CLPS beads.

The idea of antibody functionalized PS beads was enhanced further by Prasad et al [9]. Using electrophoresis, a micro-bridge of functionalized receptors was created between two sensing electrodes of the micro-electrode array (MEA). This micro-bridge acted as an amplifier for antigen-antibody binding due to the increase in sensing surface area. A rapid label free detection platform was thus demonstrated for C-reactive protein detection, which is a potential biomarker for cardiovascular diseases.

PS beads have also been used for detection of single and double stranded DNA [10]. Colored PS beads were functionalized using oligonucleotides, and then conjugated to a test strip using streptavidin – biotin chemistry. The target DNA was then applied to the functionalized surface. Then the color density of each test zones were analysed using a desktop scanner. Gold nanoparticles (AuNP) have been used for the same purpose before. The advantage of polystyrene microspheres over AuNP, arises from their superior stability in various buffers and the rapid preparation of conjugates with oligonucleotides. AuNP require a long (over 48-h) salt aging process in order to avoid aggregation at the ionic strength of the hybridization buffer. On the contrary, the oligonucleotide-decorated polystyrene microspheres are prepared in less than 3 h. In addition, polystyrene microspheres allow the development of dual-color or multicolor biosensors.

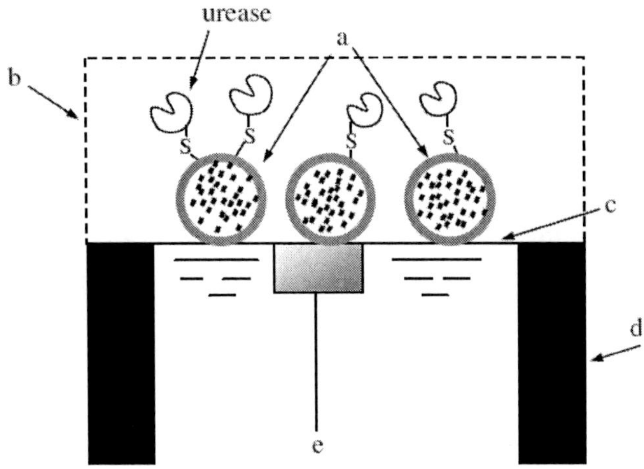

Figure 2. Cross sectional view of the urease sensor[7] fabricated by combining an ammonia electrode with urease-immobilized PMS/CLPS beads and including: (a) urease-immobilized PMS/CLPS beads, (b) dialysis membrane, (c) ammonia permeable membrane, (d) electrode body, and (e) electrolyte of the ammonia electrode.

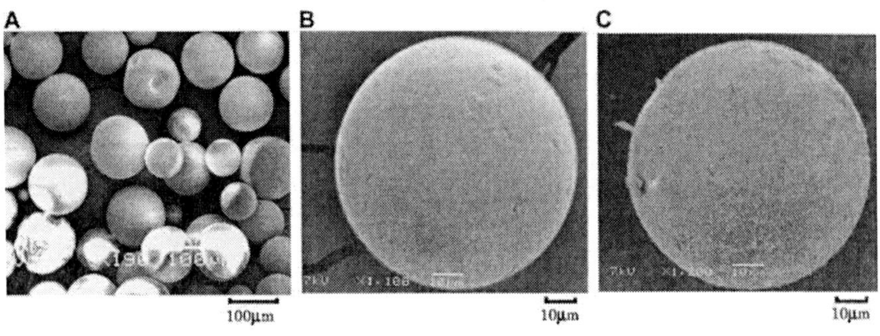

Figure 3. (A) SEM images of original CLPS beads[7], (B) magnified views of an original CLPS bead[7], and (C) a PMS/CLPS bead[7].

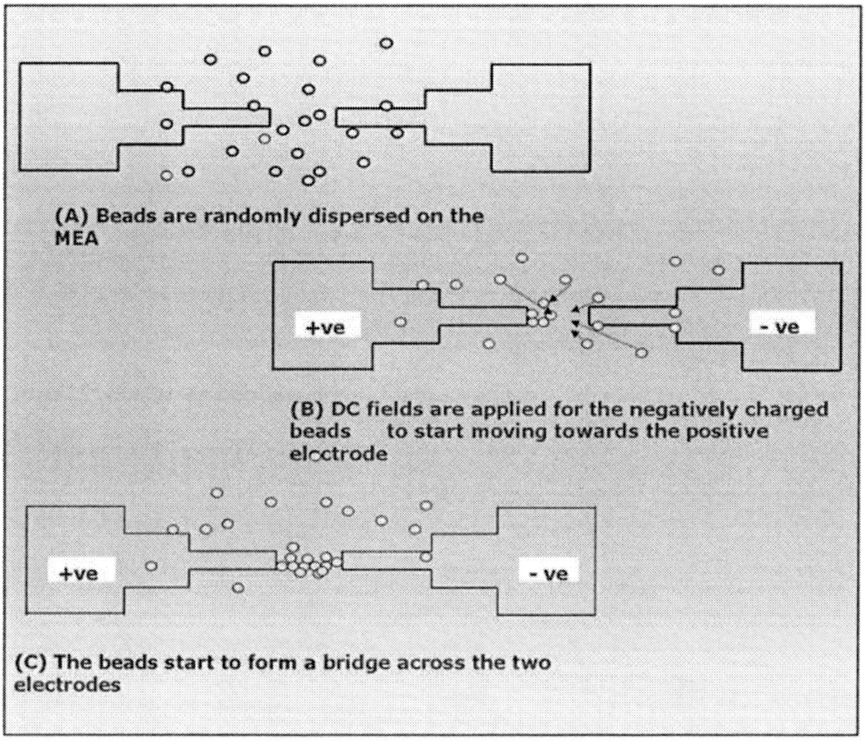

Figure 4. Formation of polystyrene micro-bridge.[9]

It is generally accepted that the biocompatibility of a material is closely connected with cell adhesion onto the surface. In the study by Lord et al [11], the cell adhesion to oxidized polystyrene and tantalum is analyzed using quartz crystal microbalance with dissipation (QCM-D) technique. Following adsorption of albumin, fibronectin or newborn calf serum onto the surfaces, QCM-D measurements showed that cells adhered and spread on the fibronectin and serum coated surfaces, while few cells adhered to the albumin coated surfaces. Cells adhered to albumin coated surfaces had a rounded morphology. The responses to fibronectin and serum pre-coated surfaces were quite different for each of the underlying substrates indicating that the process of cell adhesion and spreading elicits different responses depending on both the protein coating composition and the influence of the underlying substrate. This study proved that oxidized polystyrene is an attractive bio-compatible material which can be used in cardiovascular grafts. It also established QCM-D as a more accurate and sensitive technique as compared to fluorescence microscopy.

Another way of biomolecule detection is using Giant Magnetoimpedance (GMI) technique [12]. The target biomolecule is captured to the sensor surface using any of the previously established immobilization techniques. Cobalt ferromagnetic beads are coated with polystyrene, which are then functionalized by streptavidin. The polystyrene coating solves two purposes: to prevent their agglomeration and to facilitate their further combination with different organic molecules that work as coupling agents for target biomolecules. Depending on the distribution of the previously attached target biomolecules, the magnetic beads will attach to the biomolecules, which is then sensed using GMI based magnetic sensor.

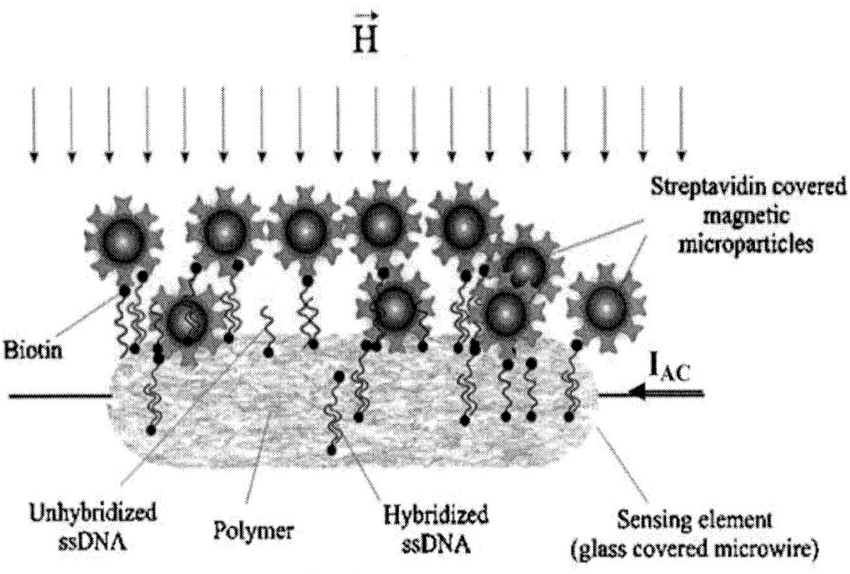

Figure 5. The principle of a GMI-based magnetic biosensor, using the ssDNA hybridization phenomenon as example [12].

Chemical Sensors

A long standing challenge in the field of the chemical detection is the development of environmental microsensors highly selective to target chemicals. The combination of suitable sensitive materials and sensing techniques is the key point for the design of microsensors-based detectors. In the field of chemical detection, materials able to interact with specific analyte by sorption phenomena are requested. Polymers are characterized by high sorption capability while they do not, usually, exhibit high selectivity. However, the easy processing techniques and the possibility to be deposited on several substrates make them reliable materials to be employed in the design of chemical detection sensors [13].

A variety of polymers have found applications in various gas sensors. The sample molecules are drawn into sensor using sampling techniques such as headspace sampling, diffusion methods, bubblers or pre-concentrators. The sample is drawn across the sensor array and induces a reversible physical and/or chemical change in the sensing material, which causes an associated change in electrical properties, such as conductivity. Each "cell" in the array can behave like a receptor by responding to different samples to varying degrees. These changes are transduced into electrical signals, which are preprocessed and conditioned before identification by a pattern recognition system. This entire process can be compared to the functionality of a mammalian nose (Figure 6). The concept was hence given the name 'e-nose (electronic nose)'. [14]

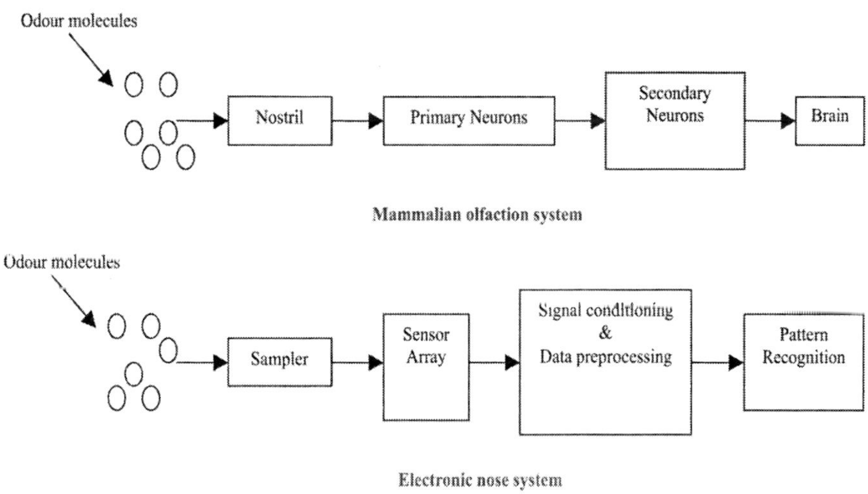

Figure 6. Comparison of the mammalian olfactory system and the e-nose system [14].

The sample molecules induce various physical changes into the sensor film material, and depending on each transduction mechanism, various sensing techniques can be employed. They are summarized in table 3 below.

Table 3. Physical changes in the sensor active film and the sensor devices used to transduce them into electrical signals

Physical changes	Sensor devices
Conductivity	Conductivity sensors
Mass	Piezoelectric sensors
Optical	Optical sensors
Work function	MOSFET sensors

Typical structures for a gas sensor using conductivity and SAW (surface acoustic wave) techniques are shown in figure 7 below.

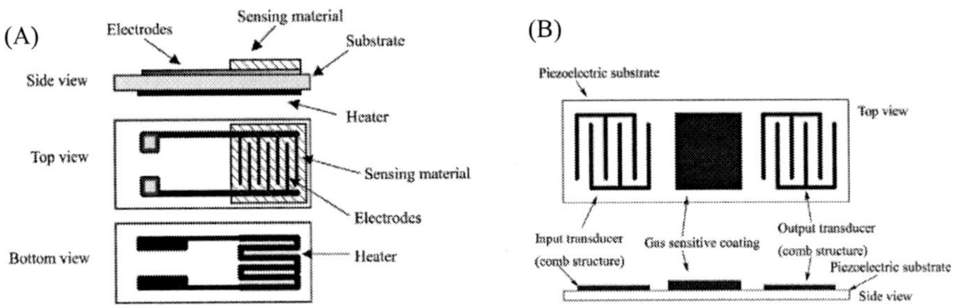

Figure 7. (A) Typical structure of a conductivity sensor[14]. (B) SAW sensor [14].

Syndiotactic polystyrene (sPS) in the semicrystalline δ form exhibits high sorption capacity and an improved selectivity towards low molecular weight substances, mainly chlorinated and aromatic, if compared to other polymers. In fact, it is characterized by the presence of a nanoporous crystalline phase which presents a regularly spaced array of nanocavities of well-defined size and shape, where penetrant molecules can be hosted.

An optoelectronic approach based on the use of fiber optic technology was implemented by Giardano et al. for sensing of chloroform and toulene. This choice seemed to be suitable in light of unrivaled advantages associated to this class of sensors such as the immunity to electromagnetic interference and the dual functionalities related to their capability to serve as transducers and sensing data transportation systems. In particular, thin film of δ phase syndiotactic polystyrene was cast by dip-coating upon the tip of a standard silica fiber optic with core of 9 μm. Reflectance measurements were carried out to detect very small amounts of chemicals (chloroform and toluene) in water down to concentrations of 5ppm [13,15].

An optical oxygen sensor based on the fluorescence quenching of AlPc(OH)–PS film by oxygen was developed by Amao et al[16]. An aluminum 2,9,16,23-tetraphenoxy-29*H*, 31*H*-phthalocyanine hydroxide (AlPc(OH), figure 8)–polystyrene (PS) film was prepared and its photophysical and photochemical properties were measured. Polycyclic aromatic hydrocarbons have strong fluorescence with long fluorescence lifetimes like 200 ns, which makes them very amenable to quenching by oxygen. However, these compounds only absorb the light in the ultra-visible and near visible range and suffer from the disadvantage of small Stoke's shift (when the excited molecule returns to the ground state, it emits light, fluorescence or phosphorescence, at a longer wavelength than the original simulating light). Thus, these compounds require complicated and expensive optical instrumentation for the separation of the excitation light from the luminescence signal. On the other hand, metal porphyrins and phthalocyanines have absorption maxima in the visible region. As the absorption maxima of metal phthalocyanines are around 600 nm, a convenient light source such as an He–Ne laser, can be used for the excitation of the phthalocyanines. Among the metallo-phthalocyanines, aluminum phthalocyanine has an exceptionally long fluorescence lifetime. Thus, aluminum phthalocyanines are attractive candidates for a new optical oxygen sensing system using fluorescence quenching.

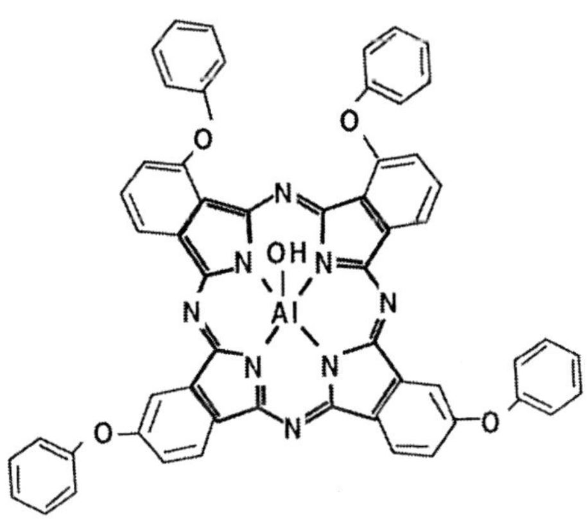

Figure 8. Structure of AlPc(OH).

AlPc(OH)–PS films were prepared by casting a solution of 30 wt.% PS in toluene and AlPc(OH) onto non-luminescence glass slides. The films were dried at room temperature and stored in the dark prior to use. Thickness of the membranes was determined by use of a micron-sensitive caliper. An excitation wavelength of 606nm was used to sense the oxygen concentration. [16]

A CO_2 sensor was developed by the same group which utilized the colorimetric change of pH indicator dye, such as thymolsulfonphthalein (thymol blue), α-naphtholphthalein and so on. In the optical sensor, the effect of surrounding humidity on the response and sensitivity is serious problem. Thus, the hydrophobic polymer film is desirable for the dye immobilization media. A hydrophobic film was created by immobilizing TPP (Tetraphenylporphryin) on polystyrene film (figure 9). [17]

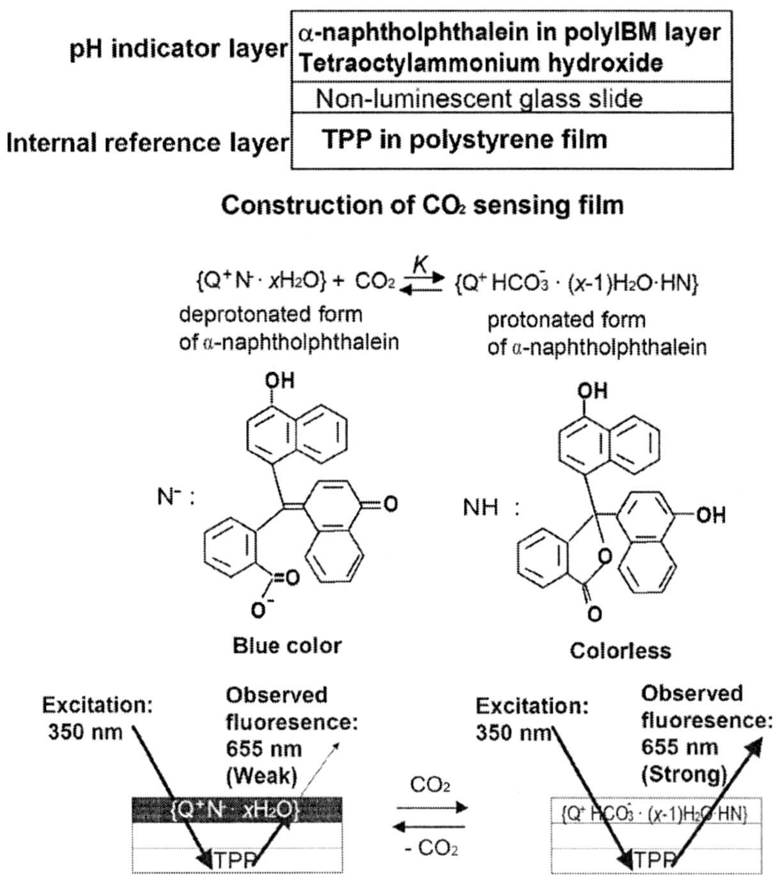

Figure 9. The construction of CO2 sensing film, the reaction process between α-naphtholphthalein and CO2, and the principle of optical CO2 sensing [17].

A NO_2 sensing device was developed by Christensen et al. based on a reversible change in d.c. conductivity in an otherwise insulating film of polystyrene upon exposure to NO_2 gas. The device consists of a polystyrene film-coated interdigitated electrode, for which several orders of magnitude increase in conductivity occurs when an NO_2/N_2 mixture, 1:10 v/v, comes into contact with the film. This rapid (2 min) change in conductivity is reversible and is not associated with any irreversible decomposition of the film. [18]

Humidity Sensors

Monitoring of moisture in air and in various gases is important in many kinds of industrial processes. During the last decade, both supply of and demand for humidity sensors have increased largely. Among the various principles applied for developing humidity sensors, the one using a resonating quartz crystal microbalance (QCM) seems still to be relatively in scant use despite its major assets. The coated piezoelectric crystal (CPC) detectors offer a low detection limit, wide measuring range and continuous mode of operation. Due to these specific features they are very appropriate for the construction of portable, simple and cheap detection units. As most of the gas sensing layers are also sensitive to water, humidity sensors are often used as part of such multi-sensing devices. A new coated piezoelectric crystal (CFC) humidity sensor was developed by Neshkova et al. based on surface chemically modified coatings of nitrated polystyrene (NPS). [19]

The CPC detector had a time response of less than 3 s, high sensitivity (up to 40Hz per %RH) and very good performance reproducibility and stability upon long-term calibration for more than 3 months. No interference with the frequency response to water vapors in the presence of corroding gases such as SO_2 and NO_2 was observed.

Solid polymer electrolytes have been developed to interact with many different gaseous substances. Humidity sensors based on solid polymer electrolytes offer many advantages such as long-term stability, reliability, ease of processing and low fabrication cost. Another humidity sensor was developed by Rubinger et al. where the changes in real part of the impedance with relative humidity was used as the transduction mechanism. [20]

The sulfonation of commercial and inexpensive polymeric materials is a powerful and versatile process, which gives the polymer a hydrophilic nature as well as proton conductivity. Sulfonated polystyrene was used as the substrate and silver electrodes were used for impedance characterization. It was demonstrated that as the relative humidity (RH) increased, the real part of the impedance dropped exponentially (figure 10).

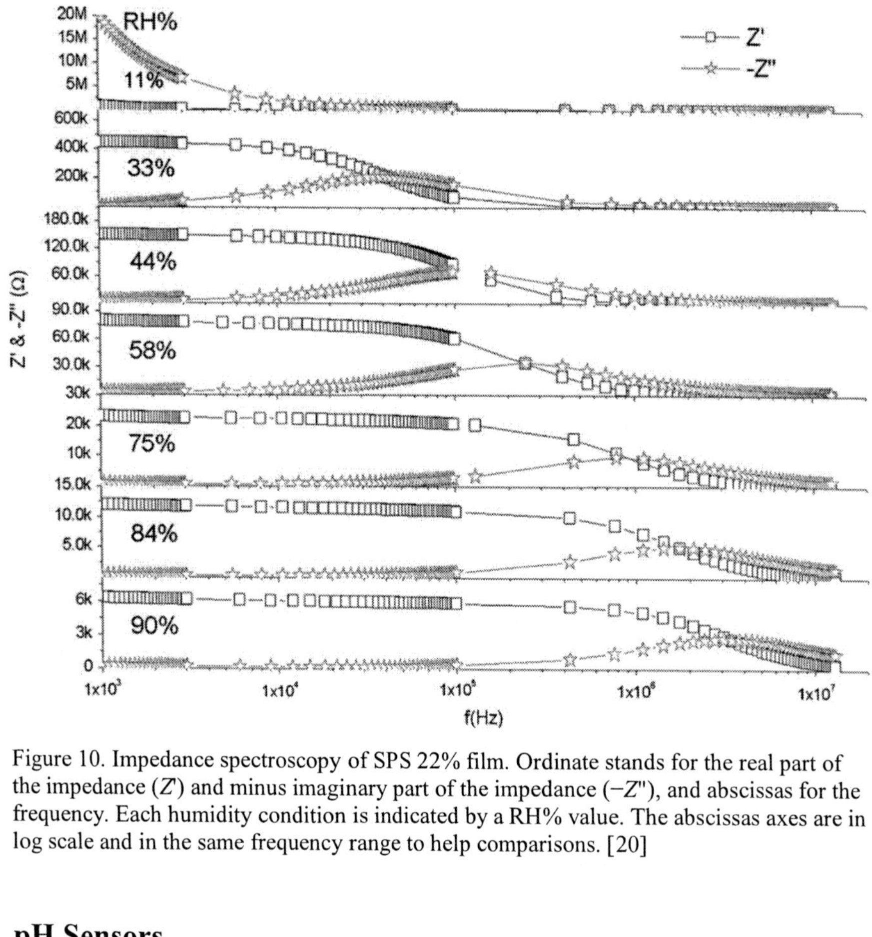

Figure 10. Impedance spectroscopy of SPS 22% film. Ordinate stands for the real part of the impedance (Z') and minus imaginary part of the impedance ($-Z''$), and abscissas for the frequency. Each humidity condition is indicated by a RH% value. The abscissas axes are in log scale and in the same frequency range to help comparisons. [20]

pH Sensors

Polystyrene has also found applications in pH sensing as demonstrated by Pan et al. pH is basically a function of the ionic concentration in the solution. It was found that depending on the concentration of ions binding to the surface of polymer beads, the beads tend to swell or shrink. Polymers with the required properties for ion sensing are not available commercially. The group had to develop the ion sensitive polymer, and they chose polystyrene because it is thermally stable, mechanically strong and is amenable to a wide variety of derivatization chemistries. Pores may be introduced by polymerizing styrene in the presence of an inert solvent, often called a porogen. This improves analyte access to the interior of the polymer.

The problem with polystyrene is that it has a glass transition temperature well above room temperature. Polystyrene based materials develop cracks and eventually fall apart when subjected to the stress of repetitive swelling and shrinking. The "toughness" of polystyrene is greatly improved by the presence of small inclusions of an elastomer. High impact polystyrene, an important commercial material, is prepared by polymerizing styrene and butadiene under conditions such that the final product has elastic polybutadiene inclusions. An alternate approach is to blend styrene-butadiene-styrene triblock copolymer with polystyrene. The butadiene blocks form a separate elastic phase improving the toughness of the polystyrene. This approach has been used to prepared a mechanically robust sulfonated polystyrene membrane. Toughness is also improved by polymerizing styrene in the presence of added triblock copolymer. [21]

The PS beads were amide derivatized and the pH sensing was demonstrated. It was shown that as the pH increased, the beads shrunk in diameter [21,22], showing a steep drop between the pH values of 6 and 8, as shown in figure 11 below.

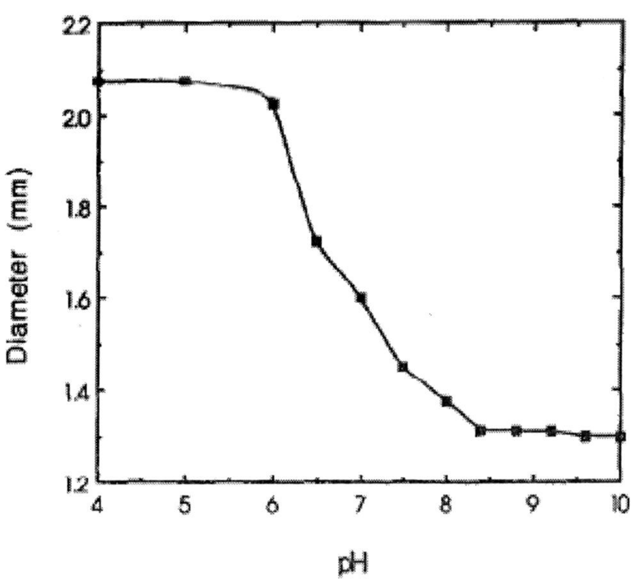

Figure 11. Diameter vs. pH at an ionic strength of 0.10 M for an aminated bead prepared with 0.8% Kraton, 5.4% DVB and 48.0% toluene.[21]

SERS Based Molecule Sensors

Surface-enhanced Raman scattering (SERS) is a phenomenon in which Raman scattering cross sections are dramatically enhanced for molecules adsorbed on nanostructured metal surfaces. Silver is the most SERS-active, followed by gold, copper, and the transition metals. In recent years, it has been reported that even single-molecule detection is possible by SERS, suggesting that effective Raman cross sections are comparable to fluorescence cross sections. Regarding the enhancement mechanism, both a long-range electromagnetic and a short-range chemical effect are thought to be simultaneously operative; factors of 8–10 orders of magnitude can arise from electromagnetic surface plasmon excitation, while the enhancement factor due to chemical effects is on the order of 10^1–10^2.

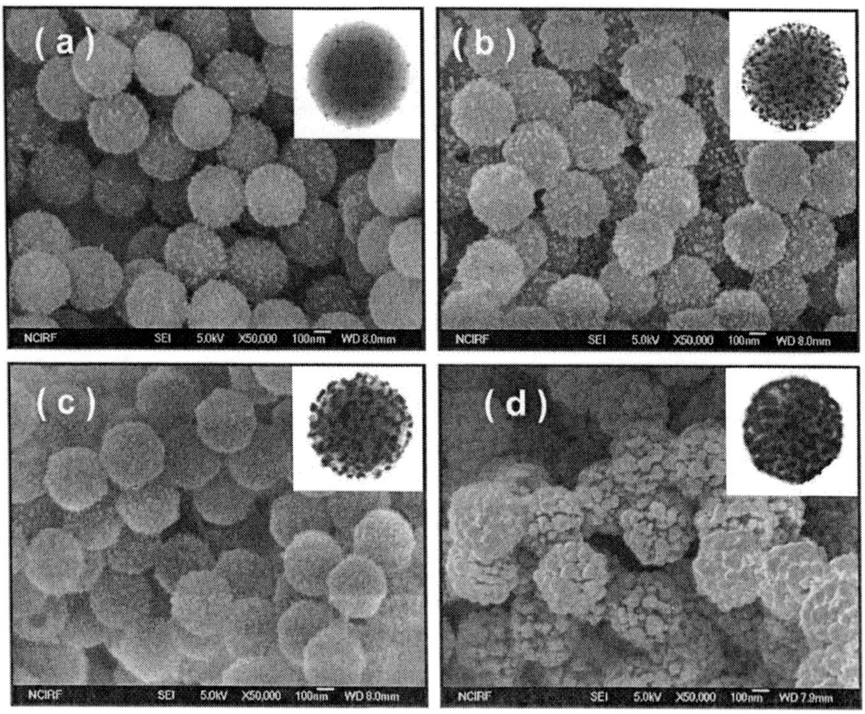

Figure 12. FE-SEM images of Ag-coated polystyrene beads prepared at molar ratios of (a) 1:0.1, (b) 1:0.25, (c) 1:0.5, and (d) 1:1 of AgNO3 and butylamine. The inset of each FE-SEM image shows the corresponding TEM image of one single polystyrene bead.

Owing to the tremendous signal enhancement that is generated, SERS has been the object of great interest in many areas of science and technology, including chemical analysis, corrosion, lubrication, heterogeneous catalysis, biological sensors, and molecular electronics. Kim et al. used silver (Ag) coated polystyrene beads as a SERS substrate (figure 12). PS was used again due to its mechanical stability. [23]

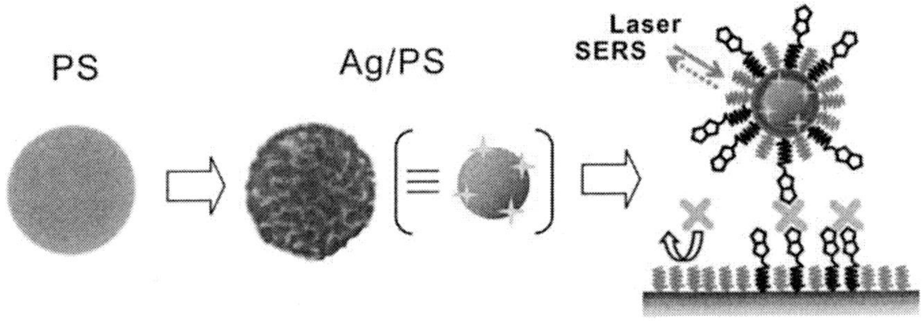

Figure 13. Fabrication of Ag-coated polystyrene (PS) bead usable as a template of biosensor operating via surface-enhanced Raman scattering; the last cartoon illustrates the interaction of biotinylated beads with other biotinylated substrates by the mediation of streptavidin molecules, labeled as "X." [23]

The beads were biotinylated and then attached to the streptavidin coated substrate (figure 13). Then optical sensing was used to detect conjugated molecules.

REFERENCES

[1] Adhikari, B. & Majumdar, S. Polymers in sensor applications. *Progress in Polymer Science* 29, 699-766 (2004).

[2] Harsányi, G. Polymer films in sensor applications: a review of present uses and future possibilities. *Sensor review* 20, 98-105 (2000).

[3] Gerard, M., Chaubey, A. & Malhotra, B. Application of conducting polymers to biosensors. *Biosensors and Bioelectronics* 17, 345-359 (2002).

[4] Nieto, A., Gaya, A., Moreno, C., Jansa, M. & Vives, J. 161-172 (Elsevier).

[5] Brizzolara, R. Patterning multiple antibodies on polystyrene. *Biosensors and Bioelectronics* 15, 63-68 (2000).

[6] Bouafsoun, A. *et al.* Electrical probing of endothelial cell behaviour on a fibronectin/polystyrene/thiol/gold electrode by Faradaic electrochemical impedance spectroscopy (EIS). *Bioelectrochemistry* 70, 401-407 (2007).
[7] Hagiwara, T., Hirata, H. & Uchiyama, S. Poly (p-maleimidostyrene) coated cross-linked polystyrene beads as novel ultrasonic irradiation resistant enzyme immobilization materials-Applications for biosensors and bioreactors. *Reactive and Functional Polymers* 68, 1132-1136 (2008).
[8] Hagiwara, T., Suzuki, I., Takeuchi, K., Hamana, H. & Narita, T. Synthesis and polymerization of N-(4-vinylphenyl) maleimide. *Macromolecules* 24, 6856-6858 (1991).
[9] Kunduru, V. & Prasad, S. Electrokinetic Formation of "Microbridges" for Protein Biomarkers as Sensors. *Journal of the Association for Laboratory Automation* 12, 311-317 (2007).
[10] Kalogianni, D., Litos, I., Christopoulos, T. & Ioannou, P. Dipstick-type biosensor for visual detection of DNA with oligonucleotide-decorated colored polystyrene microspheres as reporters. *Biosensors and Bioelectronics* 24, 1811-1815 (2009).
[11] Lord, M. *et al.* Monitoring cell adhesion on tantalum and oxidised polystyrene using a quartz crystal microbalance with dissipation. *Biomaterials* 27, 4529-4537 (2006).
[12] Chiriac, H., Tibu, M., Moga, A. & Herea, D. Magnetic GMI sensor for detection of biomolecules. *Journal of magnetism and magnetic materials* 293, 671-676 (2005).
[13] Giordano, M., Russo, M., Cusano, A., Mensitieri, G. & Guerra, G. Syndiotactic polystyrene thin film as sensitive layer for an optoelectronic chemical sensing device1. *Sensors and Actuators B: Chemical* 109, 177-184 (2005).
[14] Arshak, K., Moore, E., Lyons, G., Harris, J. & Clifford, S. Research article A review of gas sensors employed in electronic nose applications. *Sensor review* 24, 181-198 (2004).
[15] Giordano, M., Russo, M., Cusano, A. & Mensitieri, G. An high sensitivity optical sensor for chloroform vapours detection based on nanometric film of [delta]-form syndiotactic polystyrene. *Sensors and Actuators B: Chemical* 107, 140-147 (2005).
[16] Amao, Y., Asai, K. & Okura, I. Fluorescence quenching oxygen sensor using an aluminum phthalocyanine-polystyrene film. *Analytica Chimica Acta* 407, 41-44 (2000).

[17] Amao, Y. & Komori, T. Optical CO2 sensor of the combination of colorimetric change of [alpha]-naphtholphthalein in poly (isobutyl methacrylate) and fluorescent porphyrin in polystyrene. *Talanta* 66, 976-981 (2005).

[18] Christensen, W., Sinha, D. & Agnew, S. Conductivity of polystyrene film upon exposure to nitrogen dioxide: a novel NO2 sensor. *Sensors and Actuators B: Chemical* 10, 149-153 (1993).

[19] Neshkova, M., Petrova, R. & Petrov, V. Piezoelectric quartz crystal humidity sensor using chemically modified nitrated polystyrene as water sorbing coating. *Analytica Chimica Acta* 332, 93-103 (1996).

[20] Rubinger, C., Martins, C., De Paoli, M. & Rubinger, R. Sulfonated polystyrene polymer humidity sensor: synthesis and characterization. *Sensors and Actuators B: Chemical* 123, 42-49 (2007).

[21] Pan, S. *et al.* Mechanically robust amine derivatized polystyrene for pH sensing based on polymer swelling. *Analytica Chimica Acta* 279, 195-202 (1993).

[22] Zhang, L., Langmuir, M., Bai, M. & Rudolf Seitz, W. A sensor for pH based on an optical reflective device coupled to the swelling of an aminated polystyrene membrane. *Talanta* 44, 1691-1698 (1997).

[23] Kim, K., Lee, H., Park, H. & Shin, K. Easy deposition of Ag onto polystyrene beads for developing surface-enhanced-Raman-scattering-based molecular sensors. *Journal of colloid and interface science* 318, 195-201 (2008).

In: Polystyrene: Properties, Performance... ISBN: 978-1-61209-117-4
Editor: James E. Gray

Chapter 6

ELECTROSPUN POLYSTYRENE FIBERS AND SUPERHYDROPHOBIC SURFACES

Jinyou Lin[1,2,3], Bin Ding*[1,2], Jianyong Yu[2] and Gang Sun[4]

[1] State Key Laboratory for Modification of Chemical Fibers and Polymer Materials, Donghua University, Shanghai, China
[2] Nanomaterials Research Center, Research Institute of Donghua University, Shanghai, China
[3] College of Textiles, Donghua University, Shanghai, China
[4]Fibers and Polymer Science, University of California, Davis, California, USA

ABSTRACT

Polystyrene (PS), a common thermoplastic polymer made from the aromatic monomer styrene with good formability, is widely used in automotive, electrical and electronic connector systems. During the past decade, electrospinning as a versatile technique for continuous fabrication of microscale- or nanoscale-diameter fibers has attracted tremendous interests for the fabrication of ultrathin PS fibers. In this chapter, PS fibers with various morphologies including uniform fibers and beaded fibers were prepared via regulating the solution properties. Furthermore, PS fibers with the combination of micro- and nanostructured surface

* E-mail: binding@dhu.edu.cn

roughness inherently based on the inspiration of self-cleaning silver ragwort leaves were achieved directly using mixed solvents. The pure PS fibrous mats showed the superhydrophobicity with a water contact angle (WCA) of 154° and a water contact angle hysteresis of 6°. However, the practical applications for these superhydrophobic mats were significantly limited due to their inadequacy of mechanical integrity. To enhance the mechanical properties of PS mats, polyamide 6 and polyacrylonitrile fibers were blended into the PS mats via multi-syringe electrospinning technique, respectively. At the WCA of 150°, the PS fibrous mats blended with added-fibers showed a 3 times increase in the tensile strength compared with the pure PS fibrous mats.

INTRODUCTION

Electrospinning is a well recognized and effective technique to yield ultrathin fibers with diameters in the micrometers down to several nanometers from the electrostatically driven jets of polymer solutions or melts [1]. When the repulsive electric force acting on the polymer liquid overcomes surface tension at the tip of the nozzle in electrospinning, a thin polymer fluid jet will be ejected. As the charged fluid jet moves toward the collector by the electric force in the air, the solvent dries out, thus, the continuous polymer fibers are obtained as non-woven mats [2, 3].

Recently, the fabrication of fibrous mats with high surface roughness and surface-to-volume ratio via electrospinning has been widely studied [4-7]. Morphology, structure, and property of fibrous mats are controllable by adjusting the solution properties and processing parameters such as solvents, solution viscosities, applied voltage, ambient relative humidity, etc [4, 7-10]. This fabrication approach of fibrous mats is versatile and has been demonstrated to have the potential in applications such as sensors [11], filtration [12], dye-sensitized solar cells [13], tissue engineering [14], etc.

Wettability is a fundamental character of a solid surface, it is governed by both the surface chemistry or energy and the geometrical microstructures of the surface [15, 16]. Surfaces with very high water contact angles (WCAs) particularly larger than 150° are usually called superhydrophobic surfaces, which play important roles in many applications such as water repellency, biocompatibility, contamination prevention, and self-cleaning property [17-19]. In nature, lotus and silver ragwort leaves are found to have a self-cleaning ability due to their hydrophobic wax outmost layer and complicated micro- and nanostructured surfaces. So far, the techniques adopted to make artificial superhydrophobic surfaces can be simply divided into two categories: modifying a rough surface by chemicals with low surface energy [20, 21] and making a rough surface from the low surface energy materials [22, 23].

Polystyrene (PS), a common thermoplastic polymer with a low surface energy due to its CH group, has been widely studied as a typical hydrophobic polymer for superhydrophobic electrospun mat surfaces. The mimicry of the lotus-leaf-like structure by formation of microsphere/nanofiber composite mats via electrospinning a dilute PS solution has been reported by Jiang et al. [24] and Acatay et al. [25]. The WCA (160°) of the resulting PS mats is much larger than that (95°) of flat PS films. However, the practical applications for these superhydrophobic composite microsphere/nanofiber mats are significantly limited due to their inadequacy of mechanical integrity. Some clusters of microspheres separated from the mats spontaneously and floated on water droplets.

Further studies on enhancing the mechanical property of superhydrophobic fibrous PS mats have been demonstrated by Gu et al. [26] and Ding et al. [4] via electrospinning a concentrated PS solution based on the inspiration of silver ragwort leaves. The prepared PS microfibrous mats show a stable superhydrophobicity with a relatively enhanced mechanical property. Unfortunately, the improvement on the mechanical property of electrospun PS mats is very limited. The fabrication of stable superhydrophobic PS mats with high tensile strength is still a challenge. To enhance the mechanical property of fibrous PS mats, the polyamide 6 (PA6) and polyacrylonitrile (PAN) nanofibers are homogenously blended into PS mats via a four-jet electrospinning process, keeping the superhydrophobicity of the mat surface.

This chapter reviews our recent progress on the regulation of electrospun PS fiber morphology and the enhanced mechanical properties of superhydrophobic microfibrous PS mats by adding nanofibers.

FINE STRUCTURE CONTROL OF PS FIBERS

To regulate the electrospun PS fiber morphology, 30 wt%, 20 wt% and 10 wt% PS (M_n 170 000, Aldrich) solutions were electrospun from the solvent mixtures of N,N-dimethylformamide (DMF), and tetrahydrofuran (THF) (Shanghai Chemical Reagents Co., Ltd.) with various weight ratios of 4/0, 3/1, 2/2, 1/3, and 0/4, respectively. Figure 1 showed the schematic diagram of the electrospinning setup. In electrospinning, the polymer solutions was conducted at a 2 mL/h flow rate using a syringe pump (LSP02-1B, Baoding Longer

Precision Pump Co., Ltd., China), a 20 kV applied voltage (DW-P303-1ACD8, Tianjin Dongwen High Voltage Co., China) and a 15 cm tip-to-collector distance under the relative humidity of 40% at 25 °C.

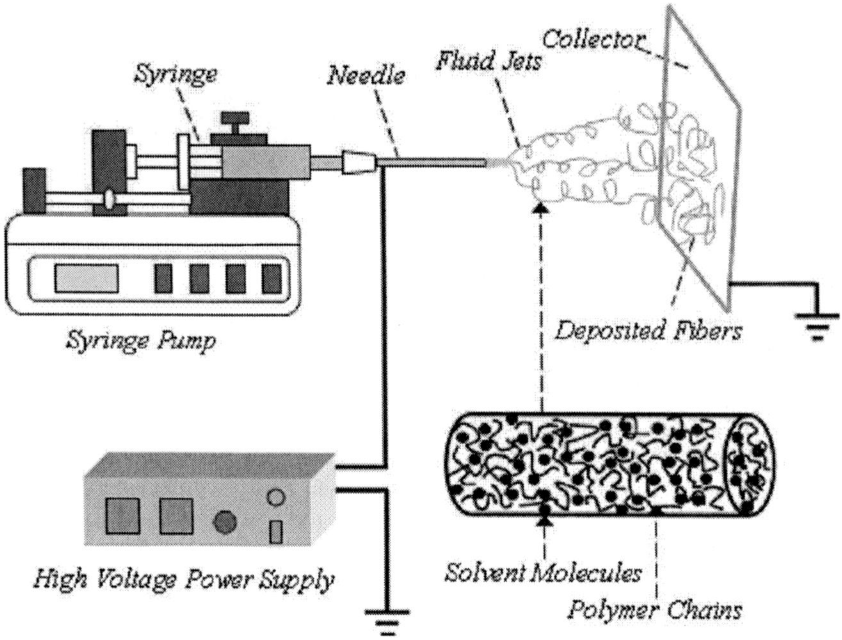

Figure 1. Schematic diagram of the electrospinning setup.

The FE-SEM (S-4800, Hitachi Ltd., Japan) images of fibers electrospun from 30 wt% PS solutions of various solvent compositions showed all fiber diameters to be in micrometer range (Figure 2) as be expected from the high polymer concentration. The average fiber diameters were 11.58, 6.17, 3.87, 6.63, and 5.78 μm from solvents at 4/0, 3/1, 2/2, 1/3, and 0/4 THF/DMF weight ratios, respectively. The fibers electrospun from THF alone displayed a ribbon-like shape with densely packed nanopores on the fiber surfaces (Figure 2a). In mixed solvent with 25% DMF, the surface nanopores disappeared but wrinkled surface emerged (Figure 2b). Fibers electrospun from the equal THF and DMF mixture appeared irregularly coarse on the surfaces (Figure 2c). With further increased DMF in the solvent mixtures, the fiber surfaces became smooth (Figures 2d,e). The presence of surface nanopores seemed to be only on fibers electrospun from the high vapor pressure THF. With added DMF, the decreased vapor pressure slowed solvent evaporation. Decreasing of solvent evaporation from the jets allowed the charged jets to remain fluid, to continue to elongate, and the collapse resulting from atmosphere pressure and electrical repulsion reduced. It has been reported that highly volatile solvent utilized in electro- spinning could create nanopores on the fibers surface and the wrinkled surface resulted from buckling instabilities during processing [8, 27].

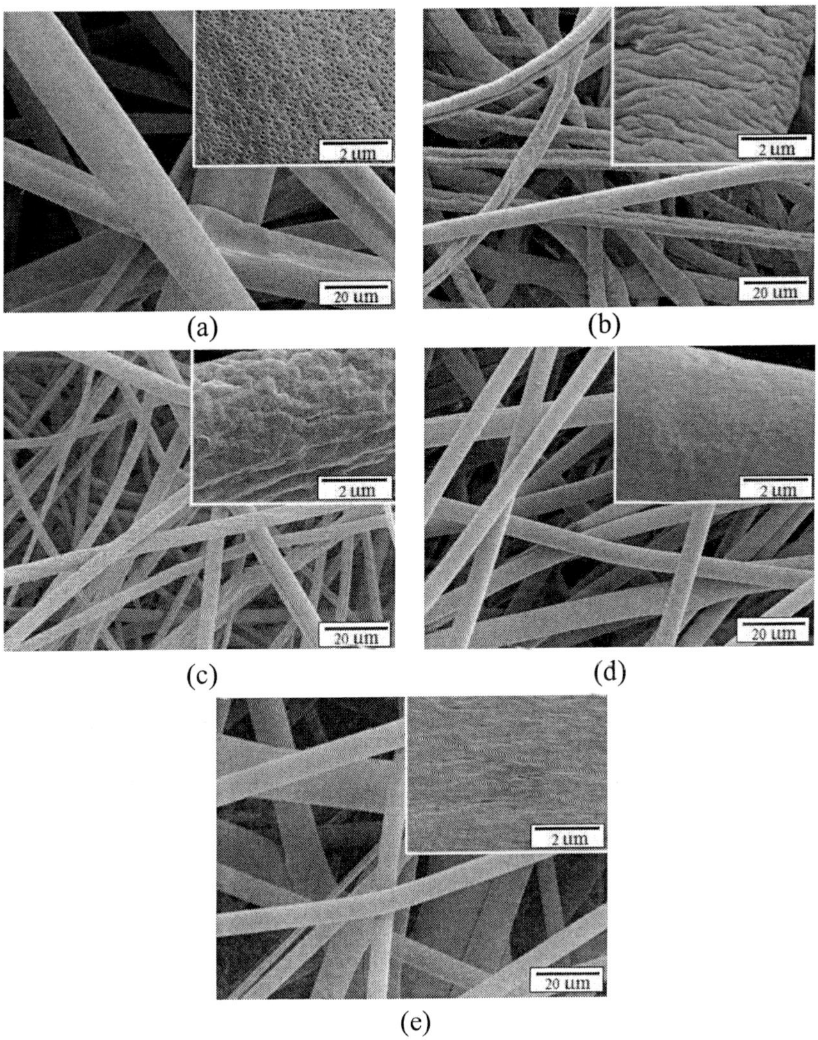

Figure 2. FE-SEM images of the PS fibers electrospun from 30 wt % PS solutions with different weight ratios of THF/DMF: (a) 4/0, (b) 3/1, (c) 2/2, (d) 1/3, and (e) 0/4 (RH, 40%).

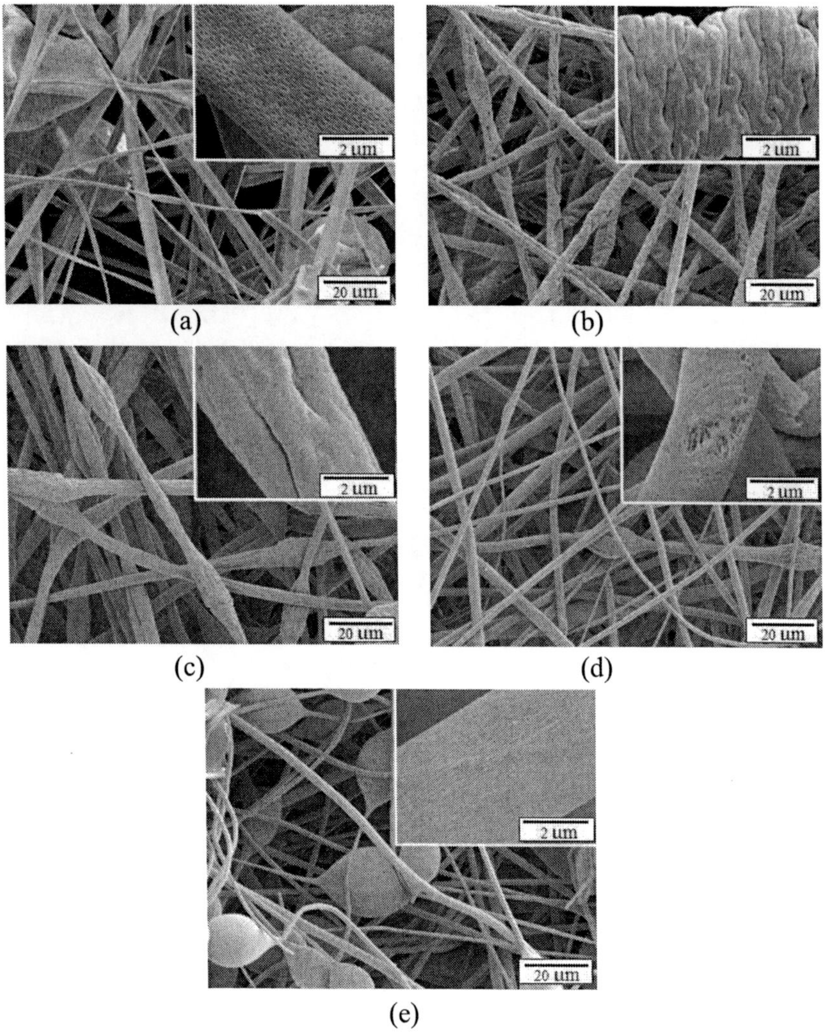

Figure 3. FE-SEM images of the PS fibers electrospun from 20 wt % PS solutions with different weight ratios of THF/DMF: (a) 4/0, (b) 3/1, (c) 2/2, (d) 1/3, and (e) 0/4 (RH, 40%).

Figure 3 presented the FE-SEM images of PS fibers electrospun from 20 wt% solutions in the same series of solvent mixtures. All PS fibers were smaller than those electrospun from a higher 30 wt % PS concentration. Changing the PS solution concentration from 30 wt% to 20 wt% led to a decrease of viscosity together with a slight increase of conductivity [7], thus the continuous fiber stretching was increased when the fluid jet was accelerated to the counter electrode giving rise to thinner fibers. Additionally, many beads emerged in the PS fibrous mats formed from DMF as shown in Figure 2e. The presence of beads was ascribed to the instability of charged fluid jets in electrospinning due to the decrease in polymer concentration and the high conductivity of the solvent [28].

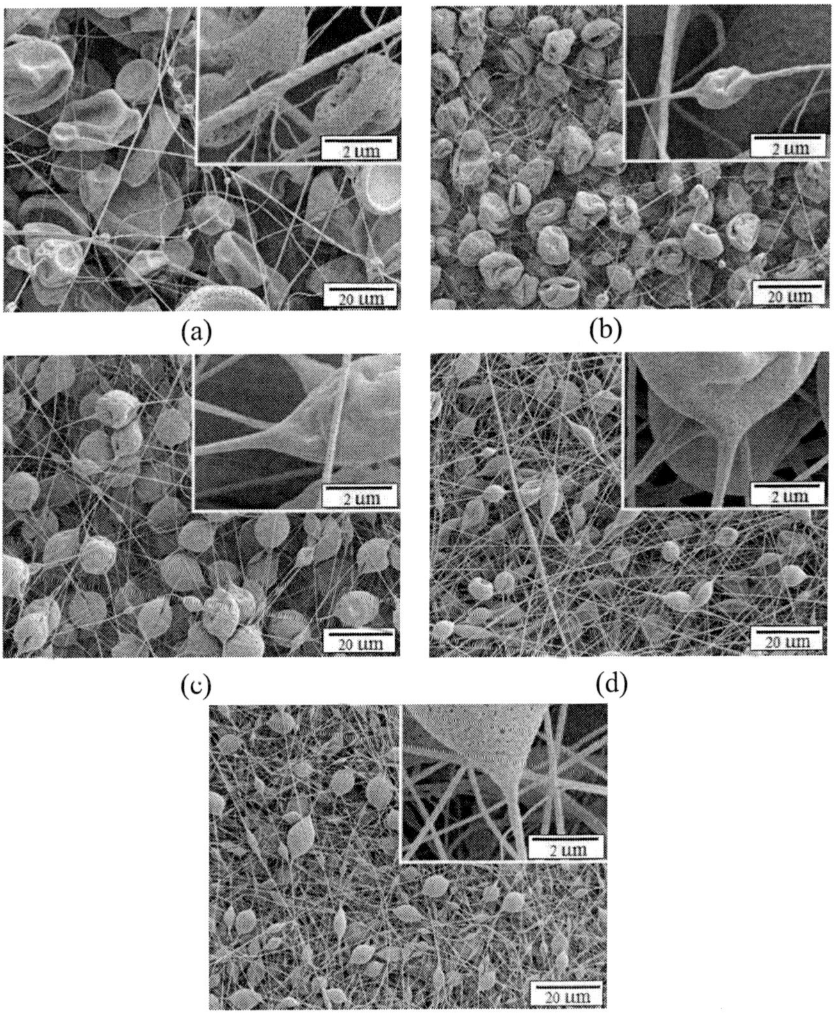

(e)

Figure 4. FE-SEM images of the PS fibers electrospun from 10 wt % PS solutions with different weight ratios of THF/DMF: (a) 4/0, (b) 3/1, (c) 2/2, (d) 1/3, and (e) 0/4 (RH, 40%).

On the other hand, we examined the fibers electrospun from 10 wt% PS solutions, and the FE-SEM images were shown in Figure 4. The interesting finding was that the fibrous mats transformed the collapsed-bead-dominant morphology into thin fibers with elliptical-bead-dominant morphology as the decreasing in vapor pressure of the solvent mixtures. Meanwhile, the size of beads was decreased. These phenomena could be attributed to the fluid jets instability resulting from the lower polymer concentration, high dielectric constant and conductivity solvent used in electrospinning [28, 29].

From the results of our study aforementioned, it was noted that various morphologies of electrospun PS fibers including uniform fibers and beaded fibers can be obtained by tuning the polymer concentration to induce the instable fluid jets formation. Further more, the surface morphology of the PS fibers can be controlled via a variation of solvent compositions.

WETTABILITY OF PS MATS

Biomimicry of the Ragwort Leaf Fibers via Electrospinning

An image of silver ragwort leaves taken by a conventional digital camera was shown in Figure 5a. The water repellency (WCA of 147°) of leaves was attributed to the hierarchical micro-/nanostructure of fibers (diameters around 6 μm) on leaves and numerous grooves (diameters around 200 nm) on fibers, as seen in Figure 5b. Based upon our previous study [4], we prepared the electrospun PS fibers (30 wt% solution) from solvents at 1/3 THF/DMF weight ratio at a 20 kV, 2 mL/h flow rate and 15 cm tip-to-collector distance under the relative humidity of 25% at 25 °C to imitate the ragwort leaf fibers.

As seen in Figure 6, the PS fibers exhibited a well-developed nanotexture rough structure on fiber surface. The formation of nanostructure on fiber surface could be explained by a rapid phase separation during the electrospinning process. Rapid evaporation of cosolvent as the fluid jet was being ejected from the tip immediately gave rise to the lowering of temperature on the ejected jet. Consequently, the fluid jet became thermodynamically unstable and phase separation took place via spinodal decomposition, which yielded polymer-rich and solvent-rich phases. The concentrated polymer-rich phase solidified quickly after phase separation and built up the matrix, whereas the solvent-rich phase formed the pores [4, 30]. The region of distribution of PS fibers ranged from 2 to 6 μm and the majority was in the range of 3-4 μm due to its high solution concentration of 30 wt%

[31]. The rough surface of the silver ragwort leaf fibers (Figure 5b) with nanometer-sized grooves along the fiber axis was successfully imitated by forming nanotextures on the surface of PS fibers.

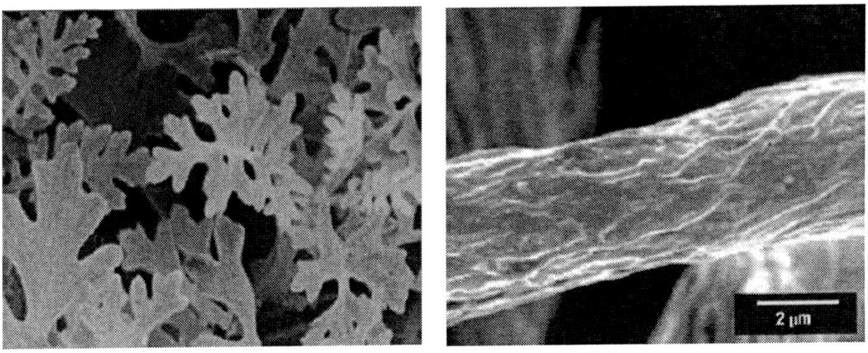

(a) (b)

Figure 5. (a) Photograph of the silver ragwort leaves. (b) SEM image of the leaf.

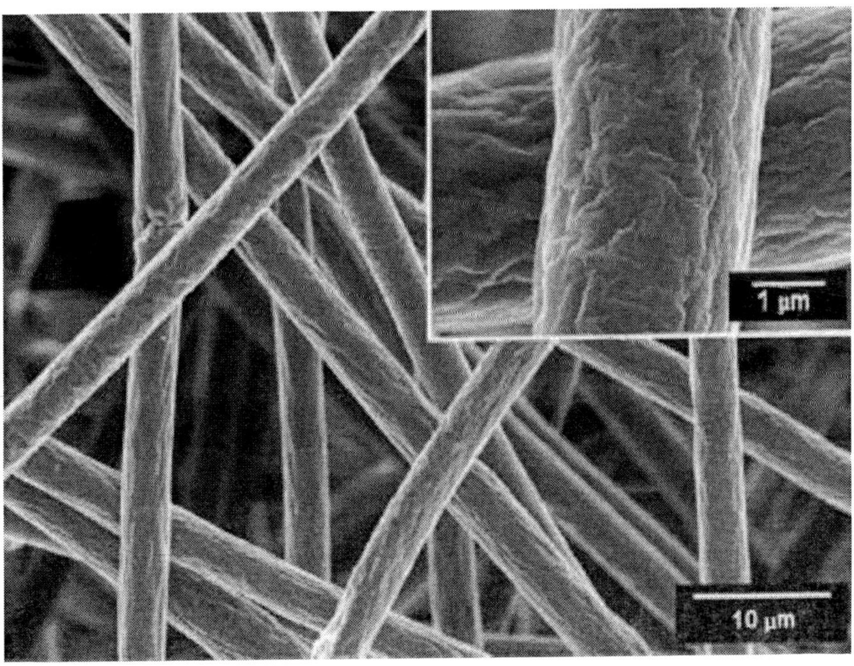

Figure 6. FE-SEM images of PS fibrous mats formed from 30 wt% PS in mixed solvents of THF and DMF with a weight ratio of 1/3 under the relative humidity of 25%.

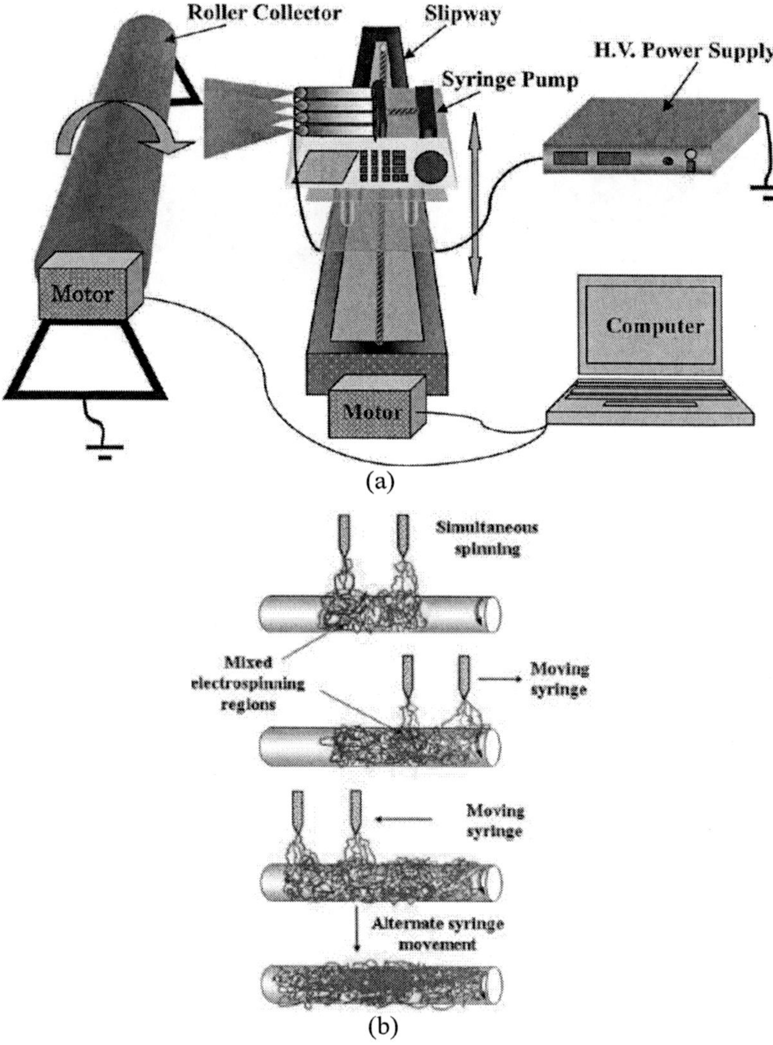

Figure 7. (a) The schematic of the four-jet electrospinning process. (b) The concept of the blending PS and added-fibers.

MODIFICATION OF THE ELECTROSPINNING PROCESS

To improve the mechanical property of fibrous PS mats, the PA6 (M_n 18 000, UBE Industries, Japan) and PAN (M_n 80,000, Institute of Shanghai Petrochemical) nanofibers were blended into PS mats via a four-jet electrospinning process as shown in Figure 7a. A grounded stainless tubular layer was used as a collector, which was covered by a piece of aluminum foil and rotated with a linear velocity of 100 m/min. A syringe pump containing four syringes was fixed to a support that could be moved with a speed of 7 m/min along a slipway. The slipway was parallel to the axis of the tubular collector. The number ratio of syringes of PS/added-fibers (PA6 or PAN) was regulated at 4/0, 3/1, 2/2, 1/3, and 0/4 to obtain the fibrous mats with various compositions of PS and PA6 or PAN. The positive electrode of a high voltage power supply was clamped to the metal needle tips, which were connected to the plastic syringe. The distance between two needle tips was 3.2 cm. The velocity of the rotatable tubular layer, the movable support, and solution feeding could be controlled by a computer. The concept of the blending PS and added-fibers were described in Figure 7b. The fibrous nonwoven mats were homogenously collected on the surface of aluminum foil and dried at room temperature under vacuum for 24 h.

Morphology of Fibrous Mats

Figure 8 showed the FE-SEM images of fibrous mats formed with various number ratios of syringes of PS/PA6. As was typical for electrospun fibrous mats, both PS and PA6 fibers were randomly oriented as porous mats with a wide fiber diameter distribution. Figures 8a-c provided the morphology of blend fibrous mats. PS and PA6 fibers in blend fibrous mats were easily distinguished because of the obvious distinction of fiber diameters. It could be seen that the PS and PA6 fibers were homogenously mixed in the blend mats. In Figure 8d, it was observed that the PA6 fibers showed a smooth fiber surface with an average fiber diameter of 132 nm. Additionally, the formation of nanowebs was observed among the PA6 fibers. The conventional electrospun PA6 fibers acted as a support for the "spider-web-like" nanowebs comprising interlinked one dimensional nanowires. The average diameter of the PA6 nanowires (14 nm) contained in this nanoweb was about one order of magnitude less than that of conventional electrospun PA6 fibers. The formation of the nanowebs was considered to be due to the electrically forced fast phase separation of the charged droplets which move at high speed between the needle tip and the collector. The phenomenon of nanoweb formation has been reported in our previous study [20, 32]. Moreover, the percentage of PA6 fiber component increased with decreasing the number ratios of syringes of PS/PA6 from 3/1 to 1/3. The utilization of rotatable collector and movable four-spinneret unit was proved to be the effective approach to produce large-scale well-dispersed blend fibrous mats.

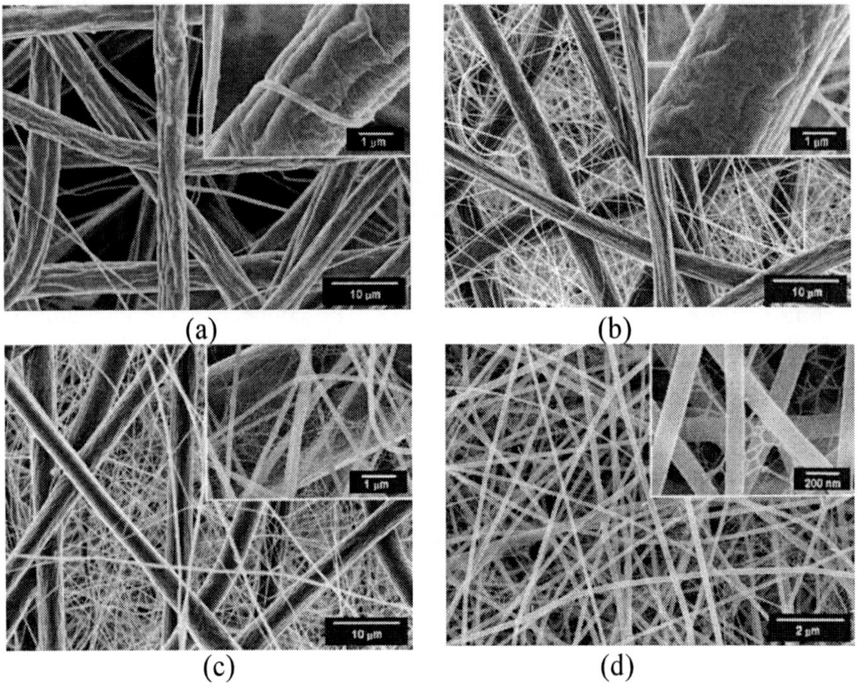

Figure 8. FE-SEM images of fibrous mats formed with various number ratios of syringes of PS/PA6: (a) 3/1; (b) 2/2; (c) 1/3; and (d) 0/4 (RH, 25%).

In another demonstration, the PAN nanofibers were blended into PS mats via a four-jet electrospinning process to improve the mechanical property of fibrous PS mats. The FE-SEM images of the blend fibrous mats of PS and PAN were shown in Figure 9. Similar result was observed that the percentage of PAN fiber component was increased with decreasing the number ratios of syringes of PS/PAN from 3/1 to 1/3. Both the PS fibers and PAN fibers were homogeneously oriented as porous mats with a wide fiber diameter distribution (Figures 9a-c). A comparison of Figure 9d to Figure 8d clearly illustrated that the average diameter of PAN nanofibers (204 nm) was larger than that of PA6 fibers due to its inherent properties of polymer [33].

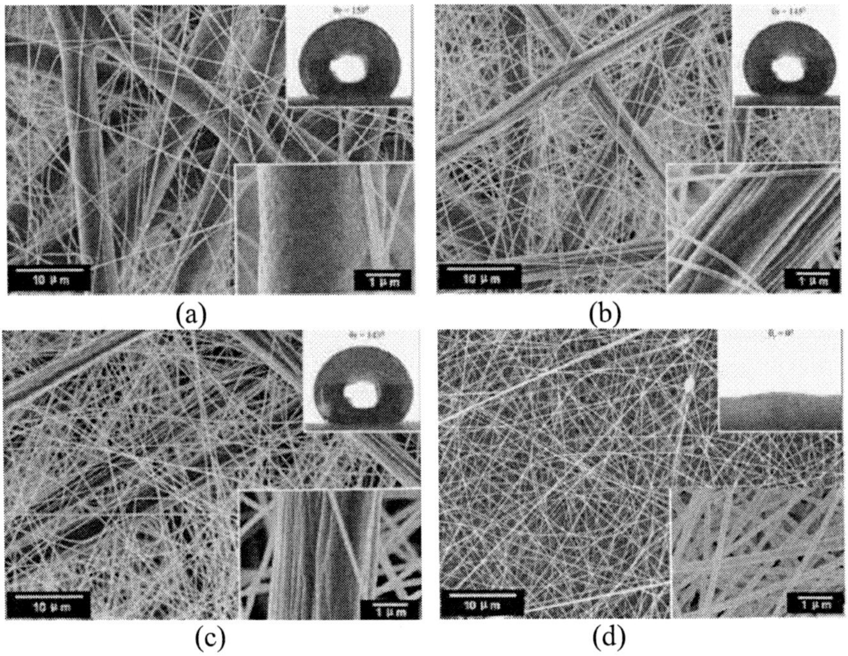

Figure 9. FE-SEM images of fibrous mats formed with various number ratios of syringes of PS/PAN: (a) 3/1; (b) 2/2; (c) 1/3; and (d) 0/4 (RH, 25%). The insets show the same sample with a larger magnifications and the profile of water droplets on the fibrous mats surface.

COMPOSITION OF FIBROUS MATS

In order to examine the composition of fibrous mats, FTIR spectroscopy (Nicolet 8700 FT-IR spectrometer) was used for the blended PS/PA6 fibrous mats formed with various number ratios of syringes of PS/PA6 and the results were shown in Figure 10. The PS fibrous mats (Figure 10a) exhibiting the absorption peaks at 3060, 3024, 1600, 1490, 1451, and 754 cm^{-1} corresponded to the phenyl (Ph) group, the peaks at 2923 and 2850 cm^{-1} to the methylene and methenyl groups, and the peak at 1028 cm^{-1} to the C-Ph bond [34]. The pure PA6 mats (Figure 10e) showed a number of FT-IR absorption features in the range 700-3400 cm^{-1} [35]. The peak at about 3300 cm^{-1} corresponded to the hydrogen bonded N-H stretch. The band at 3090 cm^{-1} was the Fermi resonance of the N-H bending vibration with the overtone of amide II (1545 cm^{-1}). The band at 2860 cm^{-1} belonged to the symmetric $-CH_2-$ stretching vibration. The strong band at 1680 cm^{-1} corresponded to the C=O stretching vibration with the amide I, and the band at about 1260 cm^{-1} arose from the C-N-H group of PA6.

The blend fibrous mats (Figures 10b-d) have IR features of both PS and PA6. In Figure 10b, the peaks at 1545, 1680, and 3300 cm^{-1} appeared in the mats formed with a number ratio of syringes of 3/1 (PS/PA6) indicating the existence of PA6 in the mats. With further decreasing the number ratios of syringes of PS/PA6 from 3/1 to 1/3, the intensities of peaks at 1545, 1680, and 3300 cm^{-1} were gradually increased showing the increase of percentage of PA6 in the blend mats. As a result, the intensities of feature peaks of fibrous mats were strongly affected by the number ratios of syringes of PS/PA6 and gradually transformed from PS to PA6 with decreasing the ratios from 4/0 to 0/4. The regular transforms indicated that PS and PA6 fibers dispersed into each other in the blend mats very well [31].

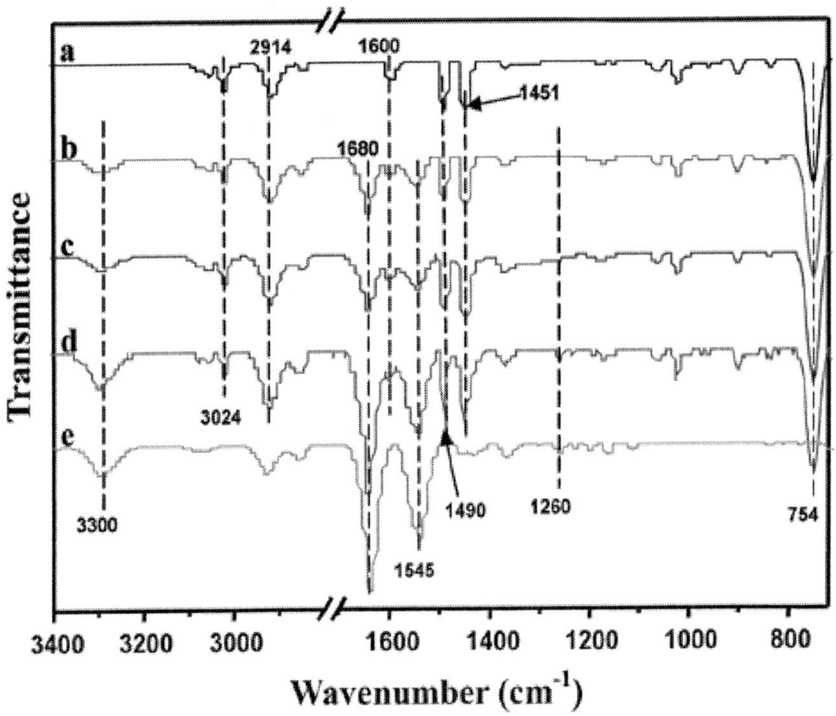

Figure 10. FT-IR spectra of fibrous mats formed with various number ratios of syringes of PS/PA6: (a) 4/0; (b) 3/1; (c) 2/2; (d) 1/3; and (e) 0/4.

HYDROPHOBICITY OF FIBROUS MATS

Figures 11a and b presented the hydrophobicity of the blend PS fibrous mats formed with various number ratios of syringes of PS/PA6 and PS/PAN, respectively. The pure PS fibrous mats showed the superhydrophobicity with a WCA (Contact Angle System OCA40, Dataphysics Co., Germany) of 154° and a water contact angle hysteresis (WCAH) of 6° (Figure 11). The superhydrophobicity of fibrous PS mat surfaces was attributed to the low surface energy of PS and the combination of hierarchical micro- and nanostructures inherent in the fibrous mats. This artificial fibrous mat achieved a stable superhydrophobic state with a WCA (154°) exceeding that (147°) of the silver ragwort leaf. It was reported that the increased surface roughness will result in an increased nominal surface energy [36]. On the PS mats surface, the roughness prevented water from contacting the PS surface completely by trapping air bubbles at the water-PS interface. The WCA of air was generally regarded as 180°. The proposed equation (1) by Cassie and Baxter described the relationship between the WCA on a flat surface (θ) and a rough surface (θ′) composed of a solid and air [15, 37, 38].

$$\cos\theta' = f_1 \cos\theta - f_2 \qquad (1)$$

In this proposed equation, f_1 and f_2 were ratios of solid surface and air in contact with liquid, respectively. Additionally, $f_1 + f_2 = 1$. Given the WCAs of the flat PS film (95°) and the fibrous PS mats (154°), f_2 was calculated to be 0.89, which indicated that the achievement of superhydrophobicity by fibrous PS mats was mainly a result of the air trapped in the rough hierarchical micro-/nanostructures of mats.

The slightly decreased surface hydrophobicity of the PS/PA6 (WCAs of 151°, 150°, and 142°) and PS/PAN (WCAs of 150°, 145°, and 143°) blend fibrous mats was found with the blend mats formed with the number ratios of syringes of 3/1, 2/2, and 1/3, respectively. The decrease of surface hydrophobicity was considered to be caused by the increased PA6 and PAN component in the mats (Figures 8a-c and 9a-c).

The WCAH was one important factor in determining surface hydrophobicity as well as the static WCA. The WCAH of PS/PA6 fibrous mats was gradually increased from 6° to 13° with decreasing the number ratios of syringes of PS/PA6 from 4/0 to 1/3 (Figure 11a), however, the WCAH of PS/PAN fibrous mats was increased from 6° to 11° at the same condition (Figure 11b). The pure PA6 mats showed a WCA of 51°, while the pure PAN mats presented a WCA of 0° indicating very good hydrophilic property due to the existence of –CN groups in PAN [39]. It was observed that the fibrous mats lost their superhydrophobicity when the number ratios of syringes of PS/added-fibers were below 2/2 and 3/1 for PS/PA6 and PS/PAN blend mats, respectively. This was explained by the PAN nanofibers had better hydrophilic than PA6 nanofibers. The large-scale fibrous mats formed with the number ratio of syringes of 2/2 (PS/PA6) and their superhydrophobicity were shown in the image of Figure 12, which was taken by a conventional digital camera.

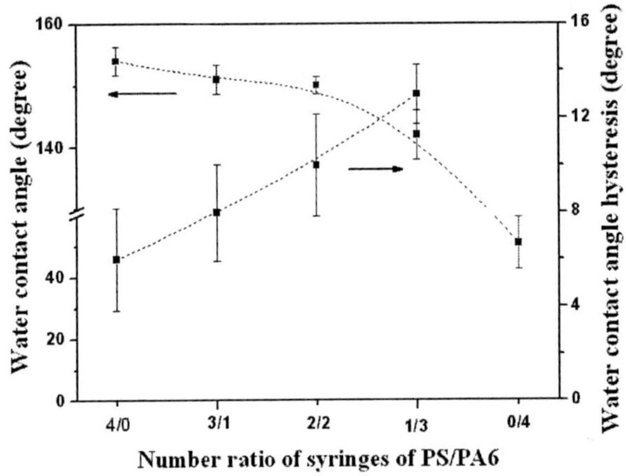

(a)

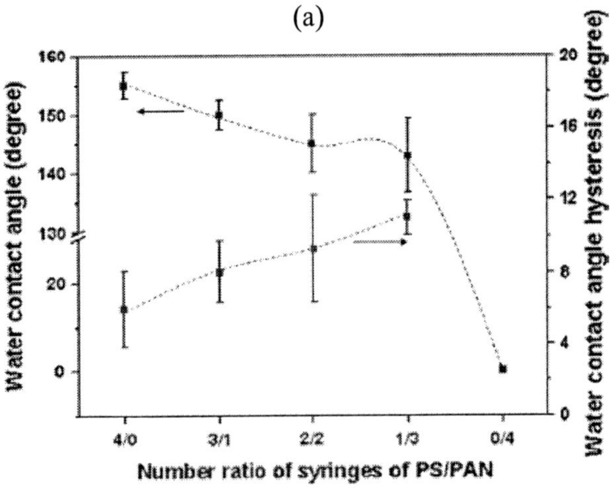

(b)

Figure 11. (a) WCAs and corresponding WCAH for the fibrous mats formed with various number ratios of syringes of PS/PA6. (b) WCAs and corresponding WCAH for the fibrous mats formed with various number ratios of syringes of PS/PAN.

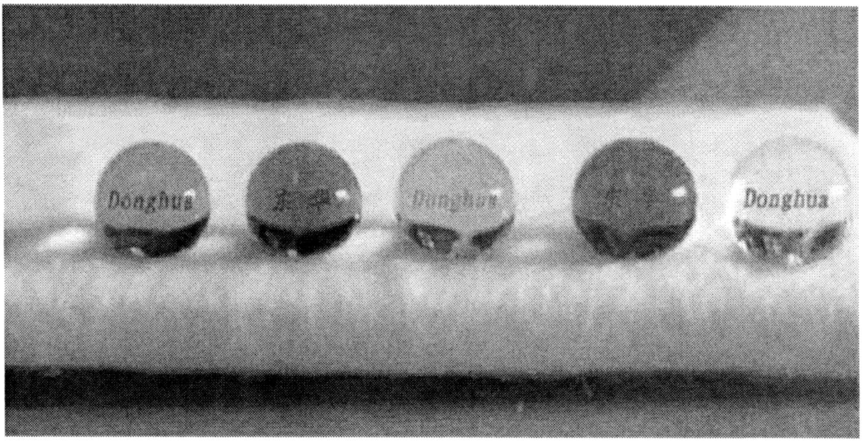

Figure 12. Several water droplets placed on the fibrous mats formed with the number ratio of syringes of 2/2 (PS/PA6) showing the superhydrophobicity.

MECHANICAL PROPERTY OF FIBROUS MATS

The mechanical properties of the fibrous mats were tested on a tensile tester (LLY-06E, Laizhou Electron Instrument Co., China) with a cross head speed of 20 mm/min. All samples were prepared in the form of standard dumbbell shape according to ASTM Standard D638 via die cutting from fibrous mats. Five specimens from each mat were tested for tensile behavior. The mechanical behavior of electrospun blend fibrous mats depends on the fiber structure, the properties of the blend components, and the interaction between each polymer fiber [40, 41]. Stress-strain behavior of fibrous mats formed with various number ratios of syringes of both PS/PA6 and PS/PAN was shown in Figures 13 and 14, respectively. The corresponding modulus, yield stress, and tensile strength of the resulting fibrous mats were shown in Figure 15. It could be observed that the pure fibrous PS mats showed the lowest modulus (150 kPa), yield stress (2.29 kPa), and tensile strength (5.45 kPa) among all fibrous mats. The low tensile strength and relatively high elongation behavior for PS mats was attributed to the nonbonded structure of PS fibrous mats (Figure 6) slipping through frictional entanglement under load.

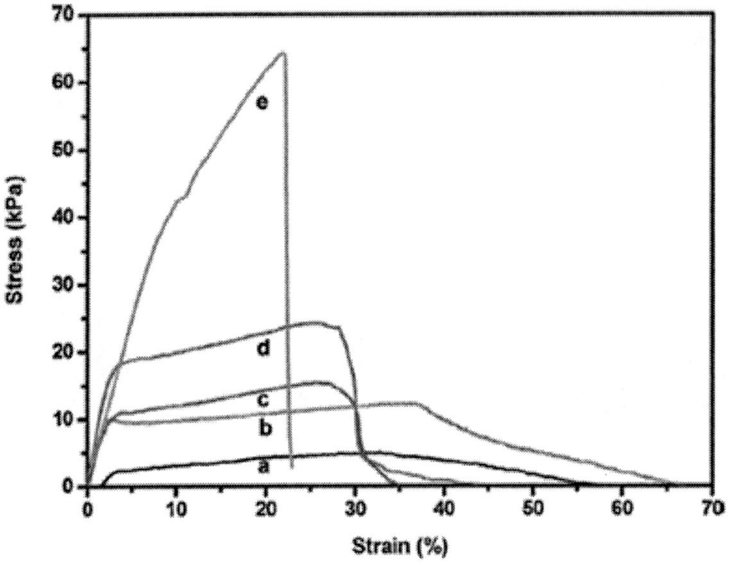

Figure 13. Stress-strain behavior of fibrous mats formed with various number ratios of syringes of PS/PA6: (a) 4/0; (b) 3/1; (c) 2/2; (d) 1/3; and (e) 0/4.

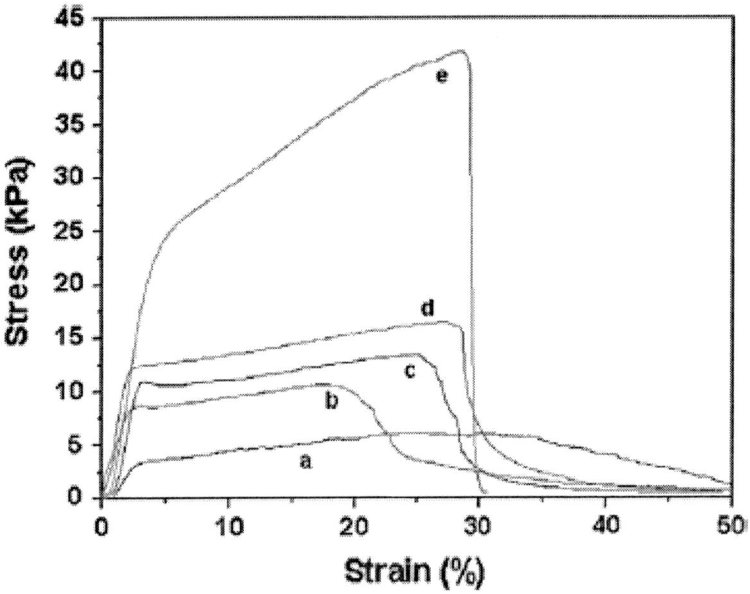

Figure 14. Stress-strain behavior of fibrous mats formed with various number ratios of syringes of PS/PAN: (a) 4/0; (b) 3/1; (c) 2/2; (d) 1/3; and (e) 0/4.

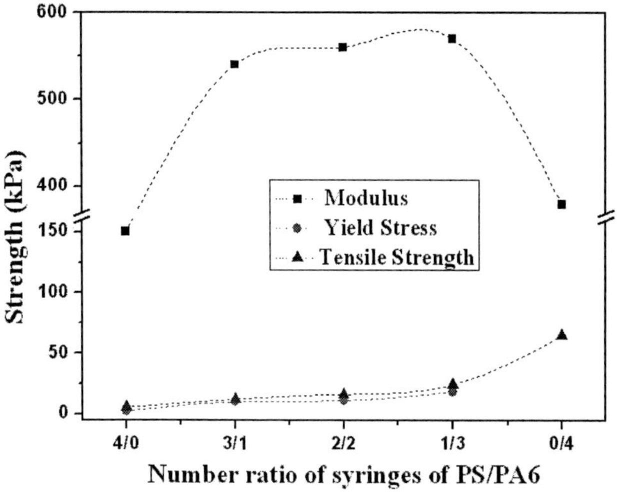

(a)

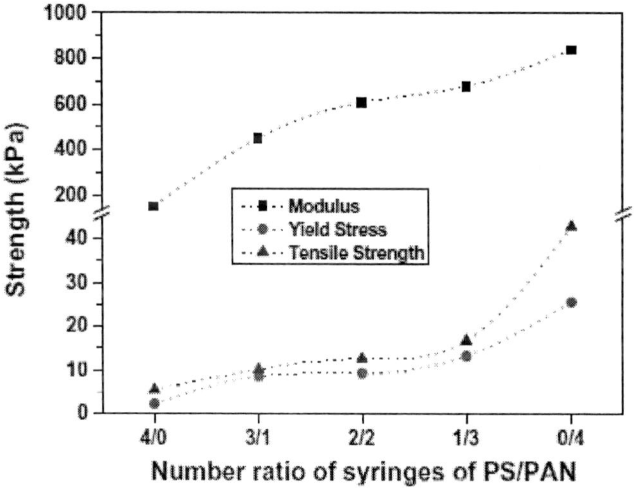

(b)

Figure 15. (a) Modulus, yield stress, and tensile strength of fibrous mats as a function of number ratio of syringes of PS/PA6. (b) Modulus, yield stress, and tensile strength of fibrous mats as a function of number ratio of syringes of PS/PAN.

The pure fibrous PA6 mats possessed the tensile strength (64.67 kPa) (Figure 15a) and a relatively low elongation at break (20.06%) (Figure 13e). Compared with pure PA6 mats, the pure PAN mats showed lower tensile strength of 42.85 kPa (Figure 15b) and much higher elongation at break 25.65% (Figure 14e). The probable reason was the difference of fiber diameters of PA6 and PAN nanofibers and inherent properties of polymers. The tensile strength of PA6 and PAN mats was nearly 12 and 8 times higher than that of PS mats, respectively (Figure 15). The point-bonded-structure among fibers in PA6 and PAN mats (Figures 8d and 9d) could be the important factor for its high tensile strength. Such point-bonded-structure was caused by the physical interdiffusion among individual electrospun fibers due to the residual solvent left after electrospinning [42].

As seen in Figure 15, the modulus, yield stress, and tensile strength of PS mat were significantly increased with adding PA6 and PAN nanofibers component compared with those of pure PS mats. The tensile strength of the blend fibrous mats was increased from 5.45 to 24.09 kPa on decreasing the number ratios of syringes of PS/PA6 from 4/0 to 1/3 (Figure 15a). For PAN nanofibers, the tensile strength of the blend fibrous mats increased from 5.45 to 16.75 kPa at the same conditions (Figure 15b). In the blend fibrous mats, PA6 and PAN nanofibers component formed a distinct elastic region before the break of mats. The network structure including a large number of entanglement and point-bonded structures among fibers (Figures 8 and 9) became the effective points for hindering the slip of PS fibers by providing local physical or frictional entanglements; such hindrance to fiber slip can gave rise to load transfer to stiffer PA6 and PAN fibers, respectively.

Additionally, the utilization of a rotatable collector and a movable four-spinneret unit made the jet move in an expanding spiral manner between the tip and the collector, and finally formed the entanglement network structures in blend fibrous mats. Consequently, the mechanical properties of blend fibrous mats were significantly enhanced with increasing the content of added-fibers (PA6 or PAN) fibers. Moreover, it also indicated that the blend fibrous mats have a good dispersibility of PS and added-fibers fibers to lead uniform mechanical properties for each mat. The tensile strength of the blend fibrous mats was increased 3 and 2 times compared with that of the pure PS mats at the critical jet ratio of 2/2 (PS/PA6) and 3/1 (PS/PAN), respectively, indicating that the PA6 nanofibers were much better suitable for improving the mechanical properties of superhydrophobic blend PS fibrous mats.

CONCLUSION

Various morphologies of electrospun PS fibers including uniform fibers and beaded fibers were successfully prepared by tuning the polymer concentration to induce the instable fluid jets formation. The solvent composition was proved to be a key factor for the control of the surface morphology of the PS fibers. The large-scale superhydrophobic fibrous mats with enhanced mechanical properties were also fabricated through the blending of biomimetic superhydrophobic PS fibers and added-fibers with high tensile strength via a four-jet electrospinning process. The obtained results indicated that the superhydrophobicity of mats was ascribed to the hierarchical structured PS fibers and the enhanced mechanical property of mats was caused by the added-fibers. The hydrophobicity and mechanical properties of fibrous mats could be regulated by tuning the number ratios of syringes of PS and added-fibers. The tensile strength of the blend fibrous mats increased 3 and 2 times compared with that of the pure PS mats showing a WCA of 150° can be achieved at the critical jet ratio of 2/2 (PS/PA6) and 3/1 (PS/PAN), respectively. Moreover, the good dispersibility of PS and added-fibers in the blend mats was observed from the results of FE-SEM, FT-IR, and mechanical properties, indicating that the adopted four-spinneret electrospinning technique is an effective approach to fabricate the large-scale well-dispersed blend fibrous mats.

ACKNOWLEDGMENTS

This work is supported by the National Natural Science Foundation of China (No. 50803009 and 10872048), the "111 Project" (No. 111-2-04 and B07024), the Shanghai Committee of Science and Technology (No. 10JC1400600), the National Basic Research Program of China (973 Program, 2011CB606100), and the Fundamental Research Funds for the Central Universities. This review collected some data from our previous publications of references 7, 31, and 33.

REFERENCES

[1] D.H. Reneker, I. Chun, *Nanotechnology* 7 (1996) 216.
[2] D.H. Reneker, A.L. Yarin, H. Fong, S. Koombhongse, *J. Appl. Phys.* 87 (2000) 4531.
[3] A.L. Yarin, S. Koombhongse, D.H. Reneker, *J. Appl. Phys.* 89 (2001) 3018.
[4] Y. Miyauchi, B. Ding, S. Shiratori, *Nanotechnology* 17 (2006) 5151.

[5] J.T. McCann, M. Marquez, Y.N. Xia, *J. Am. Chem. Soc.* 128 (2006) 1436.
[6] X. Wang, B. Ding, J. Yu, M. Wang, F. Pan, *Nanotechnology* 21 (2010) 055502.
[7] J.Y. Lin, B. Ding, J.Y. Yu, Y. Hsieh, *ACS Appl. Mater. Interfaces.* 2 (2010) 521.
[8] M. Bognitzki, W. Czado, T. Frese, A. Schaper, M. Hellwig, M. Steinhart, A. Greiner, J.H. Wendorff, *Adv. Mater.* 13 (2001) 70.
[9] S. Megelski, J.S. Stephens, D.B. Chase, J.F. Rabolt, *Macromolecules* 35 (2002) 8456.
[10] C.L. Casper, J.S. Stephens, N.G. Tassi, D.B. Chase, J.F. Rabolt, *Macromolecules* 37 (2004) 573.
[11] B. Ding, M.R. Wang, J.Y. Yu, G. Sun, *Sensors* 9 (2009) 1609.
[12] R.S. Barhate, S. Ramakrishna, *J. Membr. Sci.* 296 (2007) 1.
[13] H. Kokubo, B. Ding, T. Naka, H. Tsuchihira, S. Shiratori, *Nanotechnology* 18 (2007) 165604.
[14] S.R. Bhattarai, N. Bhattarai, H.K. Yi, P.H. Hwang, D.I. Cha, H.Y. Kim, *Biomaterials* 25 (2004) 2595.
[15] A. Nakajima, K. Hashimoto, T. Watanabe, *Monatsh Chem* 132 (2001) 31.
[16] H.Y. Erbil, A.L. Demirel, Y. Avci, O. Mert, *Science* 299 (2003) 1377.
[17] A. Nakajima, A. Fujishima, K. Hashimoto, T. Watanabe, *Adv Mater* 11 (1999) 1365.
[18] M. Miwa, A. Nakajima, A. Fujishima, K. Hashimoto, T. Watanabe, *Langmuir* 16 (2000) 5754.
[19] X.M. Li, D. Reinhoudt, M. Crego-Calama, *Chem Soc Rev* 36 (2007) 1350.
[20] B. Ding, T. Ogawa, J. Kim, K. Fujimoto, S. Shiratori, *Thin Solid Films* 516 (2008) 2495.
[21] W.L. Wu, Q.Z. Zhu, F.L. Qing, C.C. Han, *Langmuir* 25 (2009) 17.
[22] J. Bico, C. Marzolin, D. Quere, *Europhys. Lett.* 47 (1999) 220.
[23] K. Tsujii, T. Yamamoto, T. Onda, S. Shibuichi, *Angew. Chem.-Int. Edit. Engl.* 36 (1997) 1011.
[24] L. Jiang, Y. Zhao, J. Zhai, *Angew. Chem.-Int. Edit.* 43 (2004) 4338.
[25] K. Acatay, E. Simsek, C. Ow-Yang, Y.Z. Menceloglu, *Angew. Chem.-Int. Edit.* 43 (2004) 5210.
[26] Z.Z. Gu, H.M. Wei, R.Q. Zhang, G.Z. Han, C. Pan, H. Zhang, X.J. Tian, Z.M. Chen, *Appl. Phys. Lett.* 86 (2005).

[27] L.F. Wang, C.L. Pai, M.C. Boyce, G.C. Rutledge, *Appl. Phys. Lett.* 94 (2009) 151916.
[28] K.H. Lee, H.Y. Kim, H.J. Bang, Y.H. Jung, S.G. Lee, *Polymer* 44 (2003) 4029.
[29] T. Uyar, F. Besenbacher, *Polymer* 49 (2008) 5336.
[30] [C.L. Pai, M.C. Boyce, G.C. Rutledge, *Macromolecules* 42 (2009) 2102.
[31] X.H. Li, B. Ding, J.Y. Lin, J.Y. Yu, G. Sun, *J. Phys. Chem. C* 113 (2009) 20452.
[32] B. Ding, C.R. Li, Y. Miyauchi, O. Kuwaki, S. Shiratori, *Nanotechnology* 17 (2006) 3685.
[33] M. Sun, X.H. Li, B. Ding, J.Y. Yu, G. Sun, *J. Colloid Interface Sci.* 347 (2010) 147.
[34] H.L. Ge, Y.L. Song, L. Jiang, D.B. Zhu, *Thin Solid Films* 515 (2006) 1539.
[35] L.J. Li, G.S. Yang, *Polym. Int.* 58 (2009) 380.
[36] H.M. Shang, Y. Wang, K. Takahashi, G.Z. Cao, D. Li, Y.N. Xia, *J. Mater. Sci.* 40 (2005) 3587.
[37] L. Jiang, Y. Zhao, J. Zhai, *Angew. Chem.-Int. Edit.* 43 (2004) 4338.
[38] A. Cassie, S. Baxter, *Trans. Faraday. Soc.* 40 (1944) 546.
[39] G. Ma, D. Yang, J. Nie, *Polym. Adv. Technol.* 20 (2009) 147.
[40] K.H. Lee, H.Y. Kim, Y.J. Ryu, K.W. Kim, S.W. Choi, *J. Polym. Sci. Pt. B-Polym. Phys.* 41 (2003) 1256.
[41] R.B. Mathur, S. Pande, B.P. Singh, T.L. Dhami, *Polym. Compos.* 29 (2008) 717.
[42] X.H. Li, B. Ding, J.Y. Lin, J.Y. Yu, G. Sun, *J. Phys. Chem. C* 113 (2009) 20452.

In: Polystyrene: Properties, Performance... ISBN: 978-1-61209-117-4
Editor: James E. Gray

Chapter 7

COORDINATIVE CHAIN TRANSFER POLYMERIZATION: A POWERFUL TOOL FOR THE SYNTHESIS OF END-FUNCTIONALIZED SYNDIOTACTIC POLYSTYRENE

Philippe Zinck
Unité de Catalyse et Chimie du Solide UCCS, Equipe Catalyse et Chimie Moléculaire, Université Lille – Nord de France

ABSTRACT

The use of syndiotactic polystyrene for practical applications can be hampered by its brittleness, lack of adhesion and processability at high temperatures. These properties can be improved by relevant end-functionalization of the polymer. This commentary highlights the potentialities of coordinative chain transfer polymerization for this purpose. The concept of coordinative chain transfer polymerization is first introduced, followed by the use of aluminum and magnesium alkyls together with silane and borane as chain transfer agents for the synthesis of aluminum, magnesium, silicon and boron end-capped syndiotactic polystyrenes. The use of such end-capped macromolecules for additional functionalization is further discussed, with a focus on two topics:

- the formation of syndiotactic polystyrene functionalized with a variety of functions via organic transformation of the metal-carbon bond of the end-group,
- the formation of block copolymers with apolar and polar polymers.

INTRODUCTION

Syndiotactic polystyrene is a promising material that has been commercialized and widely studied by the academic community. Several drawbacks can however restrict its practical use: its brittleness, in areas where mechanical properties are important, the lack of polar groups, for adhesion and compatibility with other polymers, and the need to process syndiotactic polystyrene at high temperatures. These drawbacks can be overcome by the introduction of relevant functional groups into syndiotactic polystyrene. Coordinative chain transfer polymerization is a particularly efficient tool for the end functionalization of this polymer.

COORDINATIVE CHAIN TRANSFER POLYMERIZATION

The principle of coordinative chain transfer polymerization is represented in Figure 1. A chain transfer agent is introduced in the reactive medium at the beginning of the polymerization, generally in the form of a main group metal alkyl or hydride. The macromolecular chain grows on the catalyst, and is able to transfer to the chain transfer agent. If the chain transfer between the transition metal of the catalyst and the main group metal of the chain transfer agent is the sole termination reaction[1], all macromolecules in the reactive medium are metal-terminated via a one-pot approach. This functionalization pathway enables furthermore a fine control over the molecular weight of syndiotactic polystyrene.

Of particular interest in this context are catalyzed chain growth reactions. They involve a rapid and reversible transmetalation, occurring with high transfer efficiency[2] and in the absence of other termination pathways. All polymer chains are then bound to the main group metal in the resulting

---

[1] Other chain termination reactions such as βH transfer to the monomer or to the catalyst can occur in a competitive manner with chain transfer to the chain transfer agent.

[2] The transfer efficiency represents the ratio of metal alkyls or hydride of the chain transfer agent that are involved in the chain transfer process.

reactive medium in the absence of residual chain transfer agent (Figure 2). Further reactions can be conducted directly taking advantage of the main group metal chemistry, without any workup, and will involve the totality of the macromolecular chains present in the reactive medium. Catalyzed chain growth reactions have been reported notably for ethylene [1], isoprene [2] and styrene[3]-[4].

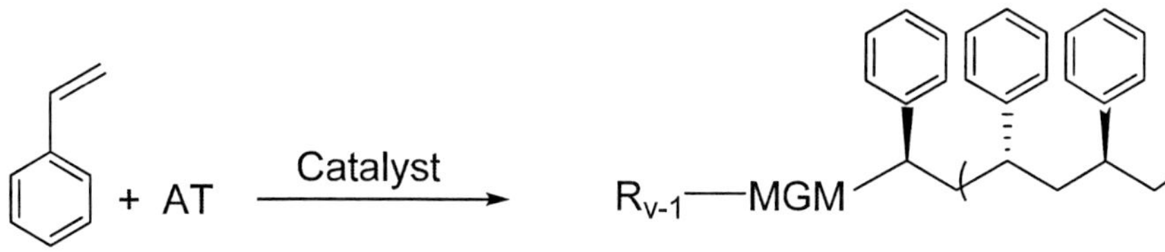

Figure 1. Coordinative chain transfer polymerization of styrene. AT represents a chain transfer agent, MGM the main group metal of the chain transfer agent, v its valency and R an alkyl and/or a hydride.

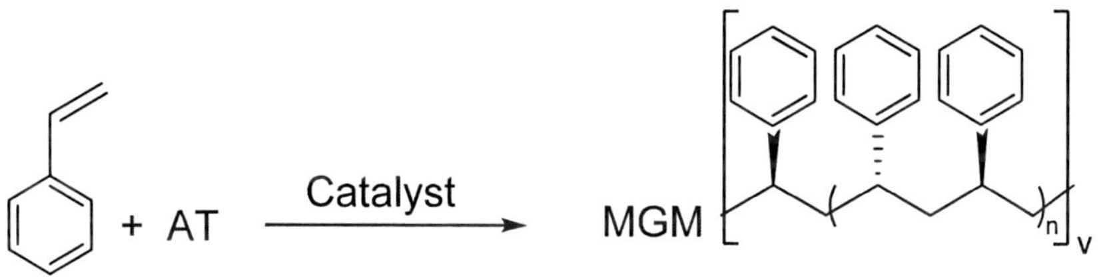

Figure 2. Catalyzed chain growth reaction. AT represents a chain transfer agent, MGM the main group metal of the chain transfer agent and v its valency.

SYNTHESIS OF MAIN GROUP METAL END-CAPPED SYNDIOTACTIC POLYSTYRENE VIA COORDONATIVE CHAIN TRANSFER POLYMERIZATION AND FURTHER DERIVATIZATION

Aluminum Based Chain Transfer Agents

Chain transfer to trialkylaluminum was reported using $CpTiCl_3$ / MAO ($Cp = C_5H_5$, MAO = methylaluminoxane) and $Ti(OEt)_4$ / MAO combined to trimethyl- and triisobutyl-aluminum [5]. The weight average molecular weight decrease is almost linear with the increase of the amount of trimethylaluminum introduced. A very sharp decrease of the weight average molecular weight is in turn observed by the addition of triisobutylaluminum, even for moderate amount. This is explained regarding the propensity of the chain transfer agent to prevent from βH abstraction. A ^{13}C NMR analysis of the polymer end groups shows that the occurrence of βH elimination tends to decrease with the amount of trimethylaluminum introduced, and disappears as triisobutylaluminum is introduced as a chain transfer agent. Chain termination occurs solely by transfer to the aluminum center in this latter case (Figure 3). It was noticed that chain transfer in the course of the polymerization can also occur on trimethylaluminum compounds contained in MAO.

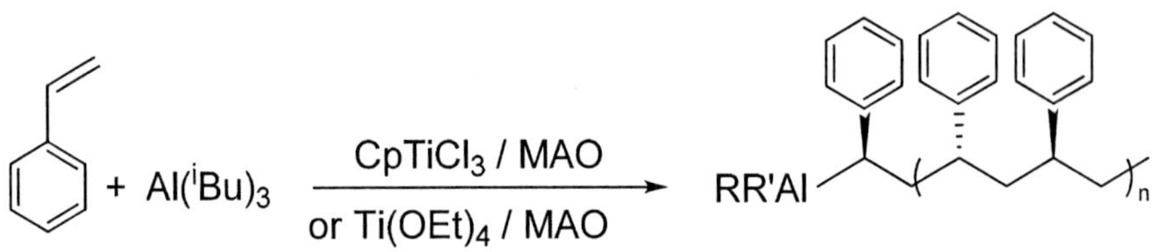

Figure 3. Syndiotactic polystyrene chain transfer to triisobutylaluminum [5]

Magnesium Based Chain Transfer Agents

Chain transfer to dialkylmagnesium was reported using the bisborohydrido halflanthanidocenes $Cp^*Nd(BH_4)_2(THF)_2$ and $Cp^*La(BH_4)_2(THF)_2$ ($Cp^* = C_5Me_5$) and *n*-butylethylmagnesium. ([3]-[4] - Figure 4). The conditions of a catalyzed chain growth reaction including full transfer efficiency were fulfilled. Magnesium terminated oligomers and polymers with well controlled molecular weight could be obtained (number average molecular weight 500 – 16 000 g/mol, polydispersity indexes 1.2 - 1.3), together with a syndiotacticity around 85%. Block copolymers with poly(1,4-*trans* isoprene) can also be synthesized.

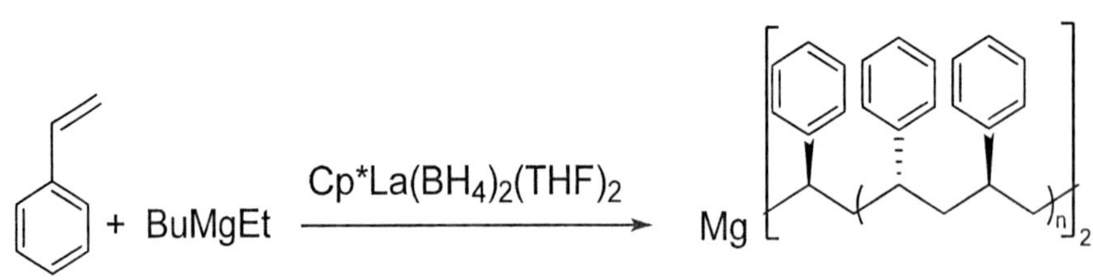

Figure 4. Neodymium catalyzed syndiotactic polystyrene chain growth on magnesium [3].

Silicon Based Chain Transfer Agents

Marks and coworkers reported the use of various silanes as chain transfer agents in the $[Cp^*TiMe_2]^+[B(C_6F_5)_4]^-$ ($Cp^* = C_5Me_5$) mediated syndiospecific polymerization of styrene ([6]-[7] - Figure 5). A 96-98% syndiotactic polystyrene where all chains are silyl capped could be obtained using $PhSiH_3$ ($Ph = C_6H_5$) as the chain transfer agent. 60 to 90% of the macromolecular chains were found to be Si terminated using $PhMeH_2Si$, *Me2H2Si*, and Et_2H_2Si. Number-average molecular weights of 40 000 to 80 000 g/mol were reported. An atactic polypropylene-*block*-syndiotactic polystyrene can be synthesized by this way using silane-functionalized polypropylene as the chain transfer agent.

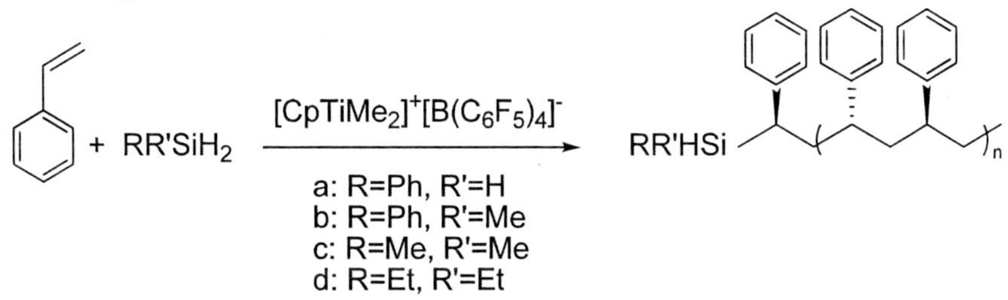

Figure 5. Syndiotactic polystyrene chain transfer to silane[6]-[7].

Vinylsilanes were also used for the syndioselective chain transfer polymerization of styrene, leading to number-average molecular weight and polydispersity indexes in the range 3000 – 20 000 g/mol and 1.6 – 2.2, respectively (Figure 6). The incorporation of the terminal vinylsilane unit at the polymer chain end occurs by 2,1-insertion. This unusual insertion pattern situates the bulky silyl functional group at a closer β-position from the active catalyst center, thus deactivating the propagating chain by a steric jam between the vinylsilane end group and the active catalyst. Subsequently, chain releasing by hydrogen addition (in the presence of H_2) or by β-elimination (in the absence of H_2) can take place, which leads to the production of end-functionalized syndiotactic polystyrene with precise controls of stereoregularity and of the location of functionality.

The resulting silane end-capped syndiotactic polystyrene can be further used for various organic transformations (Figure 7). A desilyation reaction can be conducted using trifluoroacetic acid to provide ethenyl group end-capped syndiotactic polystyrene. This functionalized macromolecule can undergo hydroboration and ozonolysis to yield hydroxyl-capped and formyl-capped syndiotactic polystyrene, respectively.

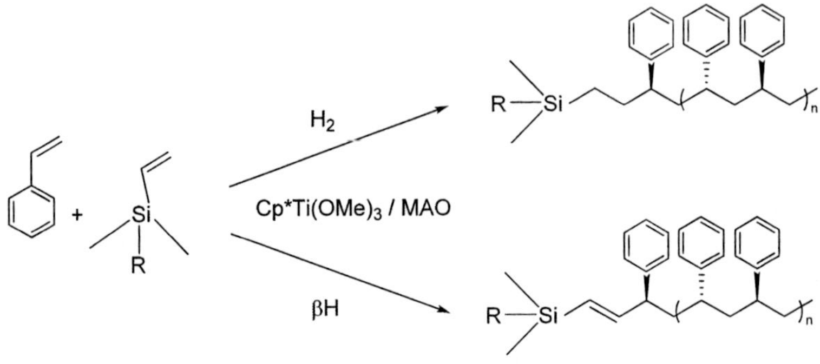

Figure 6. Syndiotactic polystyrene chain transfer to vinylsilane [8] - R = CH_3, C_6H_5

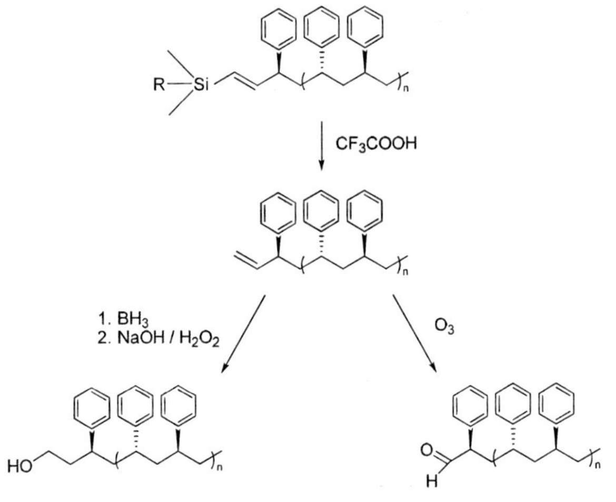

Figure 7. Derivatization of the vinylsilane end group and resulting end-functionalized syndiotactic polystyrenes[8].

Boron Based Chain Transfer Agents

Chung and coworkers used 9-borabicyclononane as a chain transfer agent in the Cp^*TiMe_3 / $B(C_6F_5)_3$ and Cp^*TiMe_3 / $B(C_{12}F_9)_3$ (Cp^* = C_5Me_5) mediated syndiospecific polymerization of styrene ([9]-[10] - Figure 8). The resulting number-average molecular weight decreased linearly with the amount of borane. Most of the polymeric chains were reported to contain a borane end group, and the syndiotacticity of the polystyrene was not affected.

The use of the more bulky tris(2,2',2'-nonafluorobiphenyl)borane instead of $B(C_6F_5)_3$ leads to a lower chain transfer activity, together with a higher activity. According to the authors, the large (tris(2,2',2'-nonafluorobiphenyl) borane $-CH_3)^-$ non-coordinated anion, located near the active half-sandwich titanocene cation, apparently does not block the active site from styrene monomer but does interfere with the bulky 9-borabicyclononane dimer.

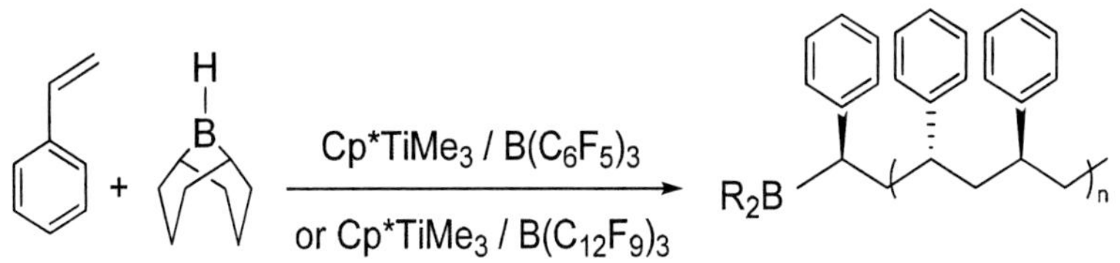

Figure 8. Syndiotactic polystyrene chain transfer to borane [9]-[10].

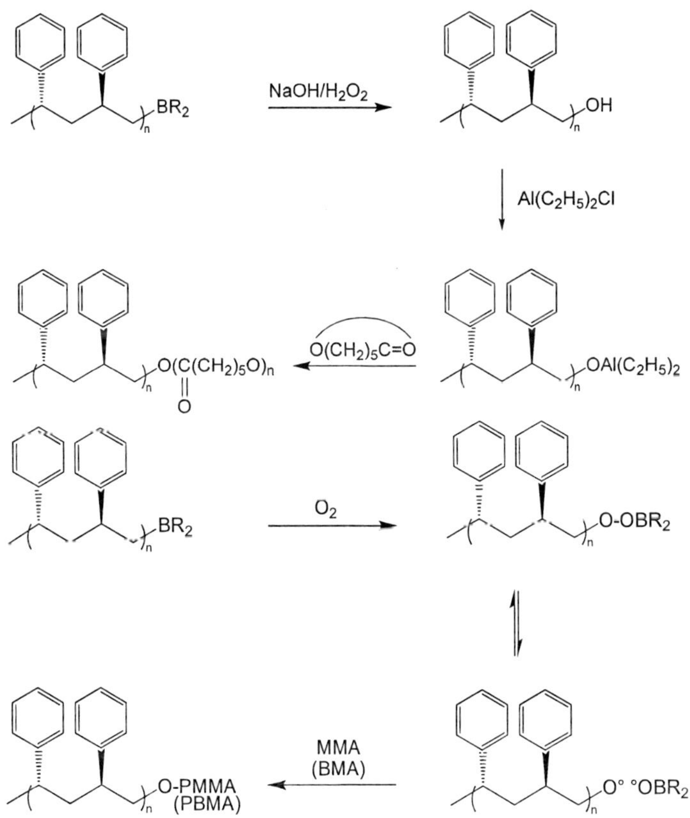

Figure 9. Derivatization of the borane end group and resulting end-functionalized syndiotactic polystyrenes [9]-[11].

The resulting borane capped syndiotactic polystyrene can be used to form (i) hydroxyl terminated syndiotactic polystyrene through NaOH/H_2O_2 mediated oxidation and (ii) block copolymers with polymethylmethacrylate and polybutylmethacrylate. The latter pathway occurs via O_2 based oxidation followed by a radical polymerization initiated through the decomposition of the peroxyde. The OH-terminated polystyrene (pathway (i)) can also react with diethylaluminum chloride to yield aluminum polystyryloxide that leads to syndiotactic polystyrene-*block*-polycaprolactone copolymers [11]. The different pathways and resulting end-functionalized syndiotactic polystyrenes are represented in Figure 9.

CONCLUSION

Coordinative chain transfer polymerization is a powerful tool for the end-functionalization of syndiotactic polystyrene. Using proper combinations of catalytic systems and chain transfer agent, aluminum, magnesium, silicon and boron end-capped syndiotactic polystyrenes can be synthesized in a one-pot procedure. These functionalized macromolecules can be further derivatized, leading to a great variety of end functionalized syndiotactic polystyrenes. This versatile strategy can also be applied to the synthesis of block copolymers with polar and apolar polymers.

ACKNOWLEDGMENTS

Dr. Andreia Valente is gratefully acknowledged for careful reading of the manuscript.

REFERENCES

[1] Pelletier JF.; Mortreux, A.; Olonde, X.; Bujadoux, K. *Angew. Chem.* 1996, 35, 1854-56.
[2] Valente A; Zinck P.; Mortreux A.; Visseaux M. *Macromol. Rapid Commun.* 2009, 30, 528-531.
[3] Zinck P.; Valente A.; Mortreux A.; Visseaux M. *Polymer.* 2007, 48, 4609-4614.

[4] Zinck P.; Valente A.; Bonnet F.; Violante A.; Mortreux A.; Visseaux M.; Ilinca S.; Duchateau R.; Roussel P. *J. Polym. Sci. Polym. Chem.* 2010, 48, 802-814.
[5] Po R.; Cardi N.; Abis L. *Polymer.* 1998, 39, 959-964.
[6] Koo K.; Marks TJ. *J. Am. Chem. Soc.* 1998, 120, 4019-4020.
[7] Koo K.; Marks TJ. *J. Am. Chem. Soc.* 1999, 121, 8791-8802.
[8] Hsiao T-J.; Tsai J-C. *J. Polym. Sci., Polym. Chem.* 2010, 48, 1690-1698
[9] Xu G.; Chung T.C. *Macromolecules* 1999, 32, 8689-8692.
[10] Chung T.C.; Xu G.; Lu Y.; Hu Y. *Macromolecules.* 2001, 34, 8040-8050.
[11] Lu Y.; Hu Y.; Chung T.C. *Polymer.* 2005, 46, 10585-10591.

INDEX

A

B

C

D

E

F

G

H

I

N

O

P

Q

R

S

T

U

V

W

X

Y